D1154223

FERROMAGNETIC-CORE DESIGN AND APPLICATION HANDBOOK

FERROMAGNETIC-CORE DESIGN AND APPLICATION HANDBOOK

M.F. "Doug" DeMaw
Member, IEEE, Graduate Engineer

Prentice-Hall, Inc., *Englewood Cliffs, New Jersey 07632*

Library of Congress Cataloging in Publication Data
DEMAW, DOUG
Ferromagnetic-core design and application handbook.

Bibliography: p.
Includes index.
1. Magnetic cores. 2. Magnetic devices.
I. Title.
TK7872.M25D45 621.34 80-16136
ISBN 0-13-314088-1

Editorial/production supervision and interior design: Nancy Moskowitz
Manufacturing buyer: Joyce Levatino

Printed in the United States of America

10 9 8 7 6 5 4 3 2 1

PRENTICE-HALL INTERNATIONAL, INC., *London*
PRENTICE-HALL OF AUSTRALIA PTY. LIMITED, *Sydney*
PRENTICE-HALL OF CANADA, LTD., *Toronto*
PRENTICE-HALL OF INDIA PRIVATE LIMITED, *New Delhi*
PRENTICE-HALL OF JAPAN, INC., *Tokyo*
PRENTICE-HALL OF SOUTHEAST ASIA PTE. LTD., *Singapore*
WHITEHALL BOOKS LIMITED, *Wellington, New Zealand*

CONTENTS

PREFACE

The work within these chapters is dedicated expressly to engineers, technicians, and college students who are presently or soon to be involved professionally with electronics. Nearly all modern circuits contain magnetic-core devices of one kind or another. It is not sufficient to the cause of good engineering practices to use inductors and transformers with magnetic cores in a casual manner. The functional parameters of toroids, rods, and pot cores are as important in a composite circuit as are the operating characteristics of the active devices being used. Therefore, more than a basic understanding of ferrites and powdered-iron components is essential today during routine design work.

Among the primary considerations when dealing with ferromagnetics are proper core selection versus operating frequency, circuit Q, power-handling capability, and physical mass. The wrong core material, or a misapplied core material, can render unusable an otherwise perfect circuit.

Emphasis has been placed on the practical aspects of magnetic core materials from low frequencies through UHF. The RF engineer or technician will find this volume invaluable in his or her daily efforts. The student will value this publication as an important textbook, and later as a standard reference for which nothing similar exists. Tedious mathematical procedures have been omitted in an effort to make comprehension more rapid and enjoyable. Equations have been used only where they are necessary to illustrate a concept or to provide a design example.

The chapters include basic theory and practical circuit examples in which toroids, rods, slugs, and pot cores are used. All of the circuits are proven ones, based on laboratory research and development by the author. The book contains myriad examples of narrow-band and broadband transformers and inductors. Ferrite loop antennas, slug-tuned inductors, RF chokes, and ferrite beads are discussed in depth, also. Magnetic-core devices are highlighted in filters, switching types of power supplies, and impedance-matching networks.

There are five appendices, which contain comprehensive design data, lists of core manufacturers, a rather lengthy bibliography, design nomographs, conversion tables, and numerous pages of component numbers and characteristics from a variety of leading manufacturers.

No credible technical author can take credit for all of the design approaches or circuits he or she commits to a text. Almost without exception the author treads unknowingly on the past work of someone else, however unintentional. Uncredited similarity between any portion of this work and that of others is unintentional and without prior knowledge on behalf of the author. Credits have been included wherever applicable.

The author acknowledges the gracious goodwill and assistance of the many manufacturers of magnetic cores discussed in this volume. The plethora of application notes and related reference data which they so readily furnished upon request contributed vastly to the relative completeness of the book. A strong vote of appreciation goes to David Boelio of Prentice-Hall, Inc. He offered more encouragement and assistance than this writer has ever experienced before. Finally, the author wishes to express his profound appreciation to Jean, his wife, for her encouragement and understanding while the book was being written.

M. F. "Doug" DeMaw

FERROMAGNETIC-CORE DESIGN AND APPLICATION HANDBOOK

1

THE BASICS OF MAGNETIC MATERIALS

In this chapter we discuss the fundamental concepts of magnetic materials and their significant characteristics and present a generalized treatment of their applications in electronics for communications engineers and technicians. The student will find this chapter a valuable base for the tutorial and practical information that is contained in subsequent sections of this volume.

One might ask: How do magnetic materials fit into the broad picture of communications electronics? The answer is simple: There are few ac circuits that do not in some manner require the use of iron, powdered iron, or ferrites. The more common applications include power-supply transformers, audio transformers, ac and RF filter inductors, broadband RF transformers, narrow-band RF transformers, and damping networks.

This book treats magnetic materials that are used as core material for the various types of electronic components just mentioned. A *magnetic core* is a configuration of magnetic material (rod, toroidal, pot core, and other physical forms), which, in its intended application, has a specific physical relationship to current-carrying conductors and whose magnetic characteristics are essential to its use.

1.1 Classifying Magnetic Materials

The most significant effect resulting from the insertion of ferrous-core material into an air-core inductor, or one that has been wound on a dielectric form, is a marked increase in inductance. The core substance used for increasing the inductance of a given coil might be classified as "hard" or "soft" magnetic material. These terms do not refer to the physical properties of the core substance. Rather, they relate to the *electrical* characteristics of the chosen material. More specifically, hard compounds have the ability to retain their magnetism after a magnetizing force has been applied to them. One application for hard-core material is found in the magnets of PM loudspeakers. The soft materials, conversely, lose their magnetism immediately upon removal of the magnetizing force. Typical uses for soft types of cores are in radio-frequency and audio-frequency transformers and reactors. This book is centered on soft-core materials. Under no circumstances should the reader assume that the soft substances discussed are lacking in hardness to the physical touch: a soft magnetic material can be extremely brittle and durable, but soft in terms of its electrical properties.

1.2 Physical and Electrical Traits

The most common physical formats for magnetic-core materials are rods; flat bars; toroids; I-E and U cores; and cup or pot cores. Figure 1-1 illustrates some of the shapes under discussion. The physical form used by the designer is dictated largely by the power level of the circuit in which the core will be used, the physical dimensions of the assembled transformer or inductor, and the fabrication time required when using one style of core as opposed to another type: Some cores lend themselves more readily to the fast production of a finished component than is possible with other core formats. Generally, magnetic core assemblies that use insulating bobbins to contain the coil or transformer windings are preferred for high-volume production. Pot cores, U cores, and I–E cores fit the foregoing description. Although toroidal and solenoidal magnetic cores can be wound with con-

Figure 1-1 Various types of magnetic-core materials.

siderable rapidity by means of machines, the process is somewhat more complicated than when dealing with plastic bobbins during mass production.

The various magnetic cores are available in a variety of sizes and compound mixes. *Laminated cores* are fashioned from various grades of steel and can be obtained in assorted widths and thicknesses. Extremely long, narrow, thin strips of steel are wound circularly and used in tape-wound toroid cores. In such an example the core is contained in a plastic housing to provide insulation between the coil winding and the core. The housing also keeps the tape core from unwinding while protecting it from moisture and chemical damage. The steel tape is "grain-oriented." This means that when the steel tape is manufactured, the rolling process is carried out to ensure that the individual grains of silicon steel (crystals) are aligned in the direction of rolling. The advantage of producing a grain-oriented steel lamination is that magnetization is much easier to realize in the direction of the grain orientation. Therefore, maximum permeability of the core will occur when the related magnetic field is parallel to the grain orientation. *Permeability* is shown symbolically by the Greek letter μ (mu) or, more specifically, $B = \mu H$ or $\mu = \mu_r \mu_0$. It is a measure of how much superior a specified material is than air as a path for magnetic lines of force. Air, the reference, has a μ of 1. Permeability is equal to the magnetic induction *(B)* in gauss, divided by the magnetizing force *(H)* in oersteds.

Tape-wound cores other than those which take toroidal form are used in a variety of modern circuits. Figure 1-2 shows the toroidal and rectangular forms under discussion. In each case a length of steel tape is wound continuously until the core has sufficient cross-sectional area to handle the amount of power required. Cores of this variety have extremely high permeability factors. Most toroidal tape-wound cores have a thin layer of paste-type insulating material between the laminations. The rectangular cores are treated with an insulating varnish between layers. This type of insulating process is necessary to prevent the formation of transverse eddy currents. When varnish is employed it also makes the laminated tape core very rigid and stable. The value of this can be seen by examining Fig. 1-3.

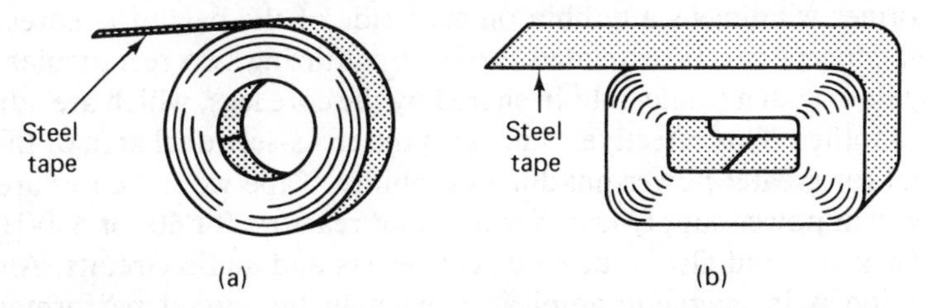

Figure 1-2 Circular and rectangular cores wound with steel tape. The type seen in (a) is toroidal, with the core in (b) shown before it is cut in half to form two C cores.

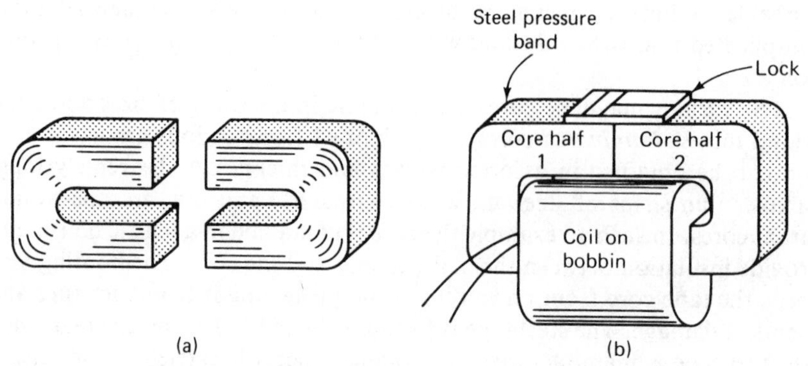

Figure 1-3 Tape-wound steel core after being cut (a) and with the bobbin in place (b). A steel band holds the core halves tightly together, as seen in (b).

Part (a) shows the evolution of a tape-wound core of the type seen in Fig. 1-2g. In Fig. 1-3a the varnished and baked core is cut in half to form a pair of grain-oriented U or C cores, as they are commonly called. Figure 1-3b shows the final assembly form of the inductor and core. One leg of each core half is inserted in the insulating bobbin on which the wire winding has been formed. A steel band is drawn very tight around the pair of U cores to maintain a minimum gap where the ends of the halves are butted together; then the band is locked in position. In some instances a metal mounting plate is affixed to the inductor or transformer assembly at the time the locking band is added: The band not only holds the core halves together tightly, but secures the mounting plate to the core material. The manufacturer may elect to encapsulate the completed unit in epoxy compound, in which case provision is made for external connection to the inductor or transformer leads by means of terminals set into the potting compound. Similarly, mounting screws or nuts can be embedded in the potting material to permit attaching the completed assembly to the chassis of the equipment in which it will be used.

The example in Fig. 1-3b can be extended to transformer use as well as to inductors. In some instances two bobbins are used to contain the transformer windings—a bobbin on each side of the pair of U cores. This general concept can be extended farther by utilizing two rectangular cores side by side, with a single bobbin shared by the core legs, which are adjacent to one another. This effectively increases the cross-sectional area of the core material for greater power-handling capability. Tape-wound cores are used primarily in power-supply transformers and reactors for 60- or 400-Hz service. They are used also in dc-to-dc converters and audio circuits. Another application is in magnetic amplifiers, wherein the circuit performance is based on the saturable-reactor concept.

1.2.1 Powdered Irons

We have established that the various physical shapes of magnetic-core materials can be applied to powdered-iron and ferrite cores as well as to those which are fashioned from steel. The choice between ferrite and powdered iron when designing a circuit is founded primarily on two basic considerations: (1) Powdered-iron cores do not saturate easily. Saturation can be defined as B_S, which is the maximum value of induction at a specified high value of field strength. This is the point at which further increase in intrinsic magnetization with an increase in field strength is minimal. (2) Ferrite cores saturate easily, thereby making them more suitable for use in circuits such as dc-to-dc converters, magnetic amplifiers, and the like. Ferrite cores offer still another advantage: The permeability of a given core can be made much higher with ferrite than is possible with powdered iron. The upper range of the latter is on the order of 90μ. With ferrite it is possible to obtain μ factors as great as 5000.

In view of the foregoing, one might surmise that there is little advantage in using powdered iron at all. Why not use ferrites for all applications in which magnetic-core material is required? Basically, the designer must effect a trade-off between high permeability and core temperature stability. As a general rule, the greater the permeability of a material, the less stable it will be at higher frequencies. The stability factor of an inductor is of paramount significance in critical circuits such as narrow-band filters, narrow-band tuned transformers, and oscillators. Ambient and RF-induced temperature changes cause the permeability to change, which in some cases can have a marked effect on stability. The higher the permeability of the core material, the more pronounced the effect. Therefore, the powdered-iron core is often the designer's preference in RF circuits that need to be relatively immune to the effects of saturation and poor stability. Iron-core inductors offer high values of Q (quality factor) and stability over a wide range of flux levels and temperatures.

Powdered-iron cores are composed of finely defined and separated particles of iron which are insulated from one another by means of the binder compound that is used in developing a particular recipe, called a *mix*. The mix is pressed into one of the various core forms (rods, slugs, and toroids) and baked at very high temperature. This technique ensures an even distribution of the powder, thereby aiding the requirement of having a relatively constant effective permeability. The useful frequency range of a mix versus the Q of the inductor wound on a core made from that mix is determined by the iron-particle makeup, the size, and the density. For example, extremely fine particles which are distributed thinly and uniformly within the binder medium permit the production of cores that are suitable for use at VHF and higher. Table 1-1 lists a popular group of powdered-iron mixes which are manufactured by Micro-Metals Corp. and sold in small quantities by Amidon Associates of North Hollywood, California. The listing is for toroid cores of various permeability factors.

TABLE 1-1 Selected powdered-iron mixes.

Material	*Color Code*	*Permeability,* μ	*Temperature Stability (ppm/°C)*	*Optimum Q Range* [a]
HA (41)	Green	75	975	1 kHz-100 kHz
HP (3)	Gray	35	370	50 kHz-500 kHz
GS6 (15)	Red and white	25	190	100 kHz-2 Mhz
C (1)	Blue	20	280	500 kHz-5 MHz
E (2)	Red	10	95	1 MHz-30 MHz
SF (6)	Yellow	8	35	10 MHz-90 MHz
W (10)	Black	6	150	60 MHz-150 MHz
IRN-8 (12)	Green and white	3	170 [b]	100 MHz-200 MHz
PH (0)	Tan	1	NC [c]	150 MHz-300 MHz

[a]Typical range rather than optimum Q range. Type HA not recommended for tuned circuits; is best suited to low-Q power transformers, noise filters, and pulse circuits.
[b]Not linear.
[c]NA, information not available.
Courtesy Amidon Associates.

1.2.2 Ferrites

The term *ferrite* can be assigned to a large number of ceramic materials that exhibit ferromagnetic properties. A *ferromagnetic compound* is one that has the capability of being magnetized to a high degree.

As is the case with powdered-iron core materials, ferrites are compounded in various mixes to achieve specific electrical characteristics. Iron oxide is combined in the binder compound with such element oxides as nickel, manganese, zinc, or magnesium. The end product is a hard, brittle substance with a smooth surface. The salient features of a ferrite core are relative ruggedness, high available μ factors, and low eddy-current losses. As mentioned during the discussion of powdered-iron core materials (sec. 1.2.1), there is a trade-off in using high-permeability ferrite cores—a degradation of stability.

An examination of the curves given in Fig. 1-4 indicates that despite the very high permeabilities attainable with ferrites, the thermal stability is not as poor as one might think. It can be seen, however, that as the permeability increases, the stability declines. The shortcoming can be minimized to a considerable degree by maintaining a fairly constant ambient temperature around the equipment area where the ferrite-loaded component is used. If we consider the portion of the curve that represents the frequency region

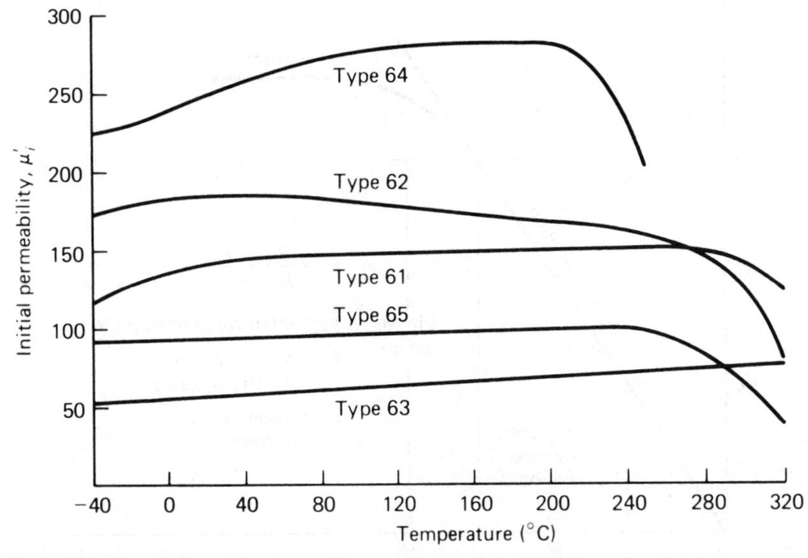

Figure 1-4 Curves that show the relationship between core temperature and initial permeability. Stability is best for cores with low permeability.

over which the core will be used (narrow), the magnitude of the temperature versus μ shift is not as dramatic.

Perhaps the most significant problem attendant to the use of ferrite cores is *saturation.* This phenomenon is defined as the state of magnetism beyond which the B–H curve levels off to a straight line. The effect, shown by the curve in Fig. 1-5, is one under which the core material cannot be magnetized further. One of the effects of saturation is called "lockup." In that state an inductor undergoes a shift in value and becomes "immobilized," so to speak. This will lead to circuit detuning and the generation of harmonics.

A ferrite-loaded inductor or transformer that is fed excessive drive power will heat considerably. In some instances the excitation can be so extreme that the core material will suffer irreparable damage. The net effect is a permanent change in core permeability. In a worst-case condition, the core can fracture and separate in many pieces.

Table 1-2 lists the principal electrical characteristics of a group of popular-size ferrite toroid cores. The data relate to materials that are manufactured by Ferroxcube Corporation and sold in small quantities by several dealers in the United States and Canada. Dimensional data are provided in inches. Although the core sizes in Table 1-2 are anything but all-inclusive for the industry, they do represent the most common size grouping used in audio and RF work. Smaller and larger toroid cores can be obtained

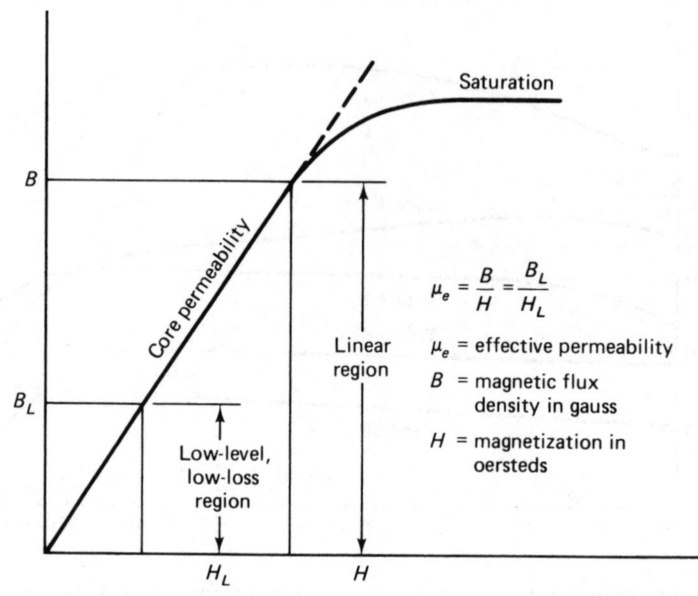

Figure 1-5 The curve shows how saturation of the core affects the linearity of the magnetic-core inductor or transformer. The B-H region is linear and the B_L-H_L region is the region of lowest loss.

TABLE 1-2

Ferroxcube core no.	*4C4 Mix*		*3D3 Mix*		*3B7 Mix*		*3C8 Mix*		*3E2A Mix*	
	A_L	μ_e	A_L	μ_e	A_L	μ_e	A_L	μ_e	A_L	μ_e
1041T060	25	125	144	725	495	2500	NA	NA	890	4495
266T125	55	125	330	750	1100	2500	NA	NA	2135	4830
768T188	70	125	415	750	NA	NA	1475	2700	2750	5000
846T250	75	125	NA	NA	NA	NA	1650	2700	3055	5000
502T300	NA	NA	NA	NA	NA	NA	1740	2700	3225	5000

Ferroxcube core no.	*A*	*B*	*C*
1041T060	0.230	0.060	0.120
266T125	0.375	0.125	0.187
768T188	0.500	0.188	0.281
846T250	0.870	0.250	0.540
502T300	1.142	0.300	0.748

Dimensions A, B and C are in inches. (25.4 × inches = mm) NA = Not Available in specified mix.

from some manufacturers on special order. Cores large enough to handle 20 kW in low-impedance balun applications are available. They are several inches in diameter.

The approximate number of wire turns for each of the cores listed in Table 1-2, versus wire gauge, are listed in Table 1-3.

TABLE 1-3 Approximate turns versus wire gauge and core type.[a]

AWG Gauge	Ferroxcube Core Number *1041T060*	*266T125*	*768T188*	*846T250*	*502T300*
16	2	5	10	25	25
18	4	10	15	35	35
20	5	15	20	45	45
22	8	20	30	55	55
24	10	25	40	75	75
26	18	30	50	95	95
28	20	40	65	115	115
30	25	50	80	145	145
32	35	65	100	180	180
34	50	80	130	240	240

[a]These data are based on the use of enameled solid-conductor wire with Formovar insulation. The number of turns specified is approximate and is intended as an aid when determining the core size versus inductance and wire size, based on the turns equation in Table 1-2.

1.3 Calculating the Coil Turns

Examination of Table 1-2 reveals that even though a particular mix of ferrite, say type 4C4, has a specific effective permeability (μ_e), the A_L factor changes with the size of the toroid core. Furthermore, the table shows that with some core materials (3E2A, for example) the μ_e value is different as the core size is increased. Some designers are not aware of this condition and are misled by the stated μ_e for a given core material, which is usually based on a fixed size of core—frequently with a diameter of 1 in. Therefore, it is an error in judgment to believe that all core sizes from a selected mix have the same μ_e, even though the basic mix recipe has a broadly based μ_i number.

The fundamental equation for determining the number of coil turns for ferrite toroid cores is

$$N = 1000\sqrt{L_{\text{mH}} \div A_L}$$

where N is the number of turns, L the desired inductance, and A_L the number assigned by the manufacturer for a given core. When A_L is not known, it can be determined by

$$A_L = \frac{L}{N^2}$$

where A_L is the desired factor, L_0 the inductance in mH for a given number of core turns (usually based on one turn), and N the number of turns used to obtain a value for L. This equation is rather awkward to use when attempting to find an A_L index for cores with low permeability, because a single turn of wire would yield an inductance in nanohenries. Accurate measurement in nH is somewhat beyond the instrumentation capability of some laboratories. When a single turn on a given core provides inductance in μH, this equation is more workable:

$$A_L = \frac{L\mu\text{H} \times 10^4}{N^2}$$

When the core material is of low μ, it is often more practical to use several turns around the core to secure the A_L index. As many as 20 turns can be useful in some instances. Thus, if a given core upon which 20 turns of wire were wound yielded a net inductance of 5 μH, the A_L would be determined by

$$A_L = \frac{5 \times 10{,}000}{400} = 125$$

Conversely, if the A_L was known to be 125 and the desired inductance was 5 μH, the number of toroid turns would be found from

$N = 100\ \sqrt{5\ \mu\text{H} \div 125} = 20$ turns where N is the unknown number of wire turns. The results are based on the coil turns being spread linearly around the entire circumference of the core material rather than being bunched up in a small area of the core. The equations assume also that a single-layer winding is employed. It is useful in some applications to arrange the coil winding to occupy approximately 180° of the toroid core. Final trimming of the inductance in critical circuits can then be done by compressing or spreading the turns on the core. This assumes that a spacing of at least one wire thickness exists between the coil turns at the offset. The amount of inductance variation will be relatively small at the lower μ_e numbers, but will increase as the μ_e of the core is made higher. Once the required inductance is effected, the turns on the core can be cemented in place by means of Polystyrene Q-Dope or a similar agent that is high in dielectric quality (Fig. 1-6).

Generally, the A_L index for ferrites is based on mH per one turn on a core, although the number of turns can be any convenient value, as stated earlier. Powdered-iron cores have their A_L indices based on μH per 100 turns. This accounts for the 10^2 (100) and the 10^3 (1000) multipliers in the two equations.

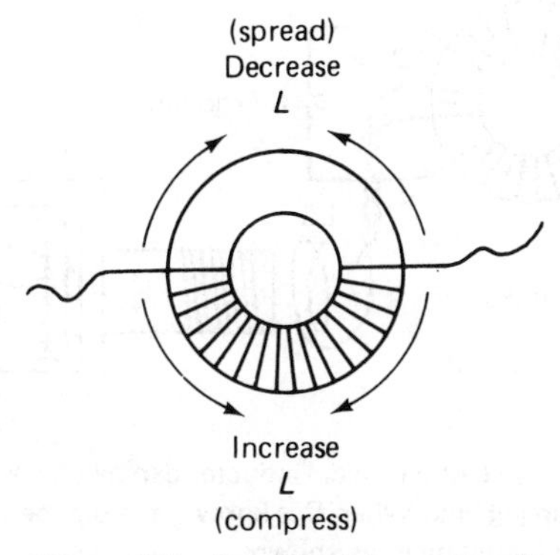

Figure 1-6 Compressing or spreading the coil turns on a toroid has a marked effect on the inductance.

1.3.1 Determining Inductance

One of the more significant benefits from the use of toroidal inductors and transformers is the inherent self-shielding characteristics of the toroidal component. This is of benefit to designers who need to engage in high-density packaging of RF circuitry without the need to include metal shield enclosures or use space-wasting separation between the inductors and near-by components. But when a laboratory inductance bridge is not available for determining A_L or making ordinary inductance measurements, a dip meter can be used if a specific technique is employed. A dip meter is a wide-range oscillator whose grid, gate, or base current is monitored by means of a dc meter. At resonances of the circuit under test, there is a sharp, well-defined reduction in indicated current. Figure 1-7 illustrates an easy method for using a dip meter with a toroidal inductor. Because the self-shielding nature of a toroid prevents coupling to a dip meter by normal means (no reading would result), it is necessary to add a temporary two- or three-turn coupling loop (L2) to the toroidal inductor, as shown. An external link, L3, is connected to L2 and the dip-meter probe coil is placed near L3. C is a close-tolerance capacitor of known value. When a dip in the meter reading is found—keeping the dip-meter probe as far from L3 as possible to obtain a shallow dip—the operating frequency of the dipper is checked against a calibrated receiver. Once f is known, the inductance can be determined by assuming that L (unknown) is equal to the reactance of C, since X_L and X_c are equal at resonance. Since the value of C in Fig. 1-7 is known, this method is straightforward. Let us assume that C in an example investigation is a 100-pF, close-tolerance silver-mica capacitor. The dip-meter output at the point where L1 and C (known) are resonant is 8.7 MHz, as observed by means of a calibrated receiver. From this we can obtain X_c for C (known):

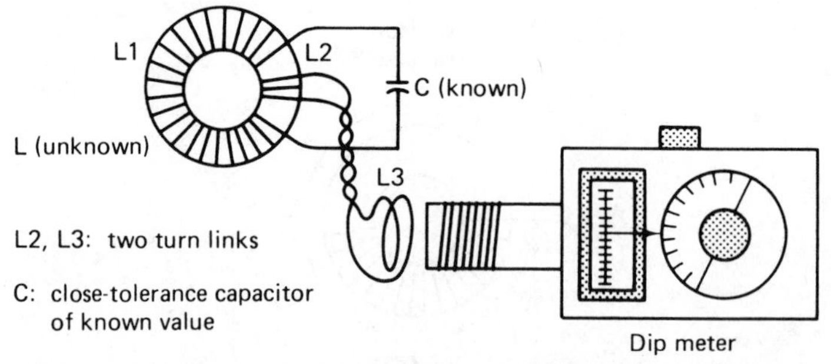

Figure 1-7 Resonance of a toroidal inductor can be checked by means of a dip meter and coupling link. When C is known, the dip meter can be used to determine the coil inductance, as shown.

$$X_C = \frac{1}{2\pi f C}$$

where X_c is the capacitive reactance in ohms, f is in MHz, and C is in μF. Thus,

$$X_c = \frac{1}{6.28 \times 8.7 \times 0.0001} = 183\Omega$$

Since $X_c = X_L$ at resonance, the inductance of L1 can be found from

$$L_{\mu H} = \frac{X_L}{2\pi f} = \frac{183}{6.28 \times 8.7} = 3.35$$

which yields an inductance value of 3.34774 μH if carried further to the right of the decimal point.

1.3.2 Circuit Q

An indication of the relative Q of a tuned circuit can be had by observing the depth of the leftward needle swing on dip meter. Tuned-circuit Q is the figure of merit. It is the ratio of reactance to resistance. The higher the tuned-circuit Q, the deeper will be the dip in meter reading. Conversely, the shallower the dip, the lower the circuit Q. When investigating the relative Q it is essential that the dip-meter probe and the tuned circuit under test be separated sufficiently to prevent overcoupling effects. A deep dip can be obtained 8 in. or more away from very high Q resonators, such as strip lines, cavities, and helical resonators. When checking toroidal inductors, as in Fig. 1-7, the dip-meter probe is more likely to be less than an inch from L3, and in some instances where the Q is medium or low, the probe may need to be inserted into L3.

When it is desired to know the actual Q_u (unloaded Q) of an air-core, ferrite, or powdered-iron-loaded inductor, a laboratory-grade Q meter, such as a Boonton 160 or Hewlett-Packard 4342A, can be used. Measurements of Q should be made at the intended operating frequency of the toroid, rod, or pot-core inductor. This rules out the use of an ordinary *RCL* type of bridge, which uses a 1000-Hz oscillator. The latter is entirely acceptable for work at audio frequencies, as would be the case with pot-core inductors or toroids intended as resonators in audio-frequency filters, and so on.

When laboratory test equipment is not available, the student or designer can apply the technique shown in Figs. 1-8 and 1-9 to obtain the unloaded Q value of a resonant circuit. The magnetic-core inductor is wound with the required number of wire turns to provide the desired inductance (L1 of Fig. 1-9). C1 is adjusted to provide resonance at the chosen f_0. C1 must be of high dielectric and mechanical quality to ensure a high Q at f_0. A signal generator is set at f_0. It should be a 50-Ω instrument, and if not, a 50-Ω pad should be inserted in the line. With an oscilloscope or RF voltmeter connected across R1, trimmers C_{in} and C_o are set for approximately equal capacitances to establish roughly 30 dB of insertion loss. Low-value trimmers in the 5- or 10-pF class are usually suitable for measurements in the MF and HF parts of the spectrum. Larger amounts of capacitance will be required for LF, VLF, and audio-frequency measurements.

Once the −30dB loss has been obtained at f_0, the signal source is varied until the half-power points on the response curve are identified. When f_0, f_1, and f_2 are known, Q_u can be obtained as shown in Fig. 1-8. Hence, if in a test circuit the value of f_0 was 2.8 MHz, f_2 was 2.79 MHz, and f_1 was 2.81 MHz, we would find Q_u by

$$Q_u = \frac{2.8}{2.81 - 2.79} = \frac{2.8}{0.02} = 140$$

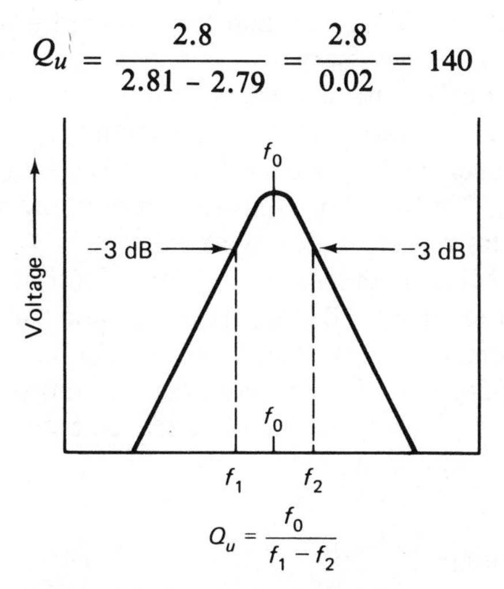

Figure 1-8 Frequency response of a resonant circuit showing the relationship between Q and the half-power points on the curve.

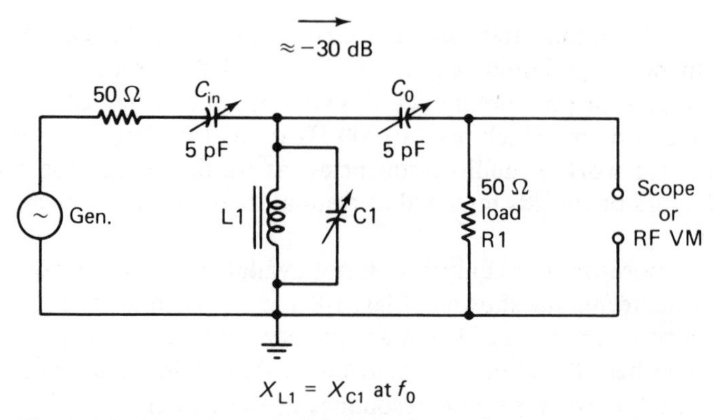

Figure 1-9 Method for checking the unloaded Q of a resonant circuit. The generator should be accurate and have suitable readout resolution for locating the -3-db points on the response curve.

where f is in MHz. The results of measurements made in the manner illustrated in Fig. 1-9 are entirely suitable for most RF design work.

When considering the available Q of an inductor, one must, of course, include the resistance of the wire, R. Thus, the larger the cross-sectional area of the conductor, the lower the resistance and the higher the coil Q. This is related to the standard equation $Q = 2\pi fL/R$, where f is in hertz, L is in henries, and R is in ohms. It can be seen from this that magnetic-core inductors are more favorably for obtaining high-Q performance than is typically possible with lumped L/C circuits which utilize inductors with many turns of small-diameter wire (Fig. 1-10). To realize Q values equivalent to those which are common to powdered-iron or ferrite-loaded inductors it would be necessary to employ large air-wound coils with substantially larger conductor diameters. With the latter it is impossible to achieve workable miniaturization. Additionally, the self-shielding characteristic of toroids and pot cores is lost if large air-wound solenoidal inductors are used. The dramatic contrast is seen in the photograph of Fig. 1-11, where the relative size of a toroidal inductor is compared to that of a section of B&W Miniductor coil stock. For identical inductance values (10μH) the Q_u of the toroid coil at 8 MHz is 350 and for the air-wound unit it is 450. It is important to point out, however, that in this example the air-wound coil is capable of handling considerably more RF power than is possible with the toroidal-wound coil. Each component has its virtues, depending upon the application.

1.4 Power Capability

When an inductor is called upon to function in a circuit where considerable power is present, the effective permeability (μ_e) of the core material will vary in a significant manner. This can be seen by referring to Fig. 1-5. In ad-

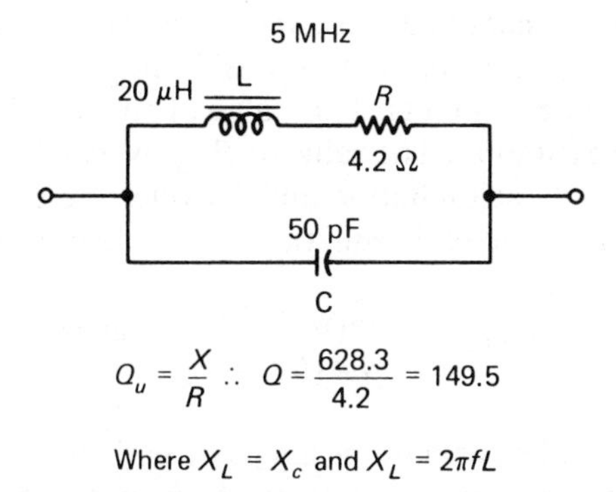

Figure 1-10 Typical resonator, showing the components *L, C,* and *R.*

Figure 1-11 How an air-wound and toroidal inductor compare when the inductance is equal.

dition to this variation in μ, the core losses will increase as a function of self-heating. This form of heating is apart from that which is brought on by environmental or ambient temperature. The self-heating effects are caused by the excitation, which in turn heats the core material and the conductors on the core. In addition to the core losses resulting from heat, the Q of the inductor can degrade markedly at peak excitation. Earlier we acknowledged the effect that this condition has on the stability of the inductor: Over a wide frequency range it can be quite poor. In a practical design application it becomes important, therefore, to acknowledge heating as an important design limit.

1.4.1 Flux Density versus μ_e

An important parameter for any magnetic material is the *flux density.* This is the density of the lines of force as measured at a cross section of their flow. To be more specific, the flux density is specified in lines/cm^2 in a magnetic circuit when measured at a given point. This is specified in units of gauss, and in the magnetic-core industry as B. B_{max} is, therefore, the maximum flux density in gauss units. This value of flux density corresponds to the peak of the ac excitation. The value of B_{max} is based on two conditions: (1) where there is only ac excitation and (2) where ac excitation is accompanied by dc voltage. The basic equation for the first condition is

$$B_{max\ (ac)} = \frac{E_{rms} \times 10^8}{4.44 f N_p A_e} \quad \text{gauss}$$

where A_e is the equivalent area of the magnetic path in cm², E_{rms} the applied voltage, N_p the number of core turns, f the frequency in hertz, and B_{max} the maximum flux density in gauss. For the second condition the equation is written as

$$B_{max} = \frac{E_{rms} \times 10^8}{4.44 f N_p A_e} + \frac{N_p I_{dc} A_L}{10 A_e} \quad \text{gauss}$$

where I_{dc} is the dc current through the winding and A_L the manufacturer's inductance for the core material in use. In this case B_{max} is classified as "total" rather than just for ac.

The curves of Fig. 1-12 show the relationship between permeability and flux density at the peak period of ac excitation for a group of core A_L values. These curves show the typical relationship of a specific shape and size of magnetic-core material which operates at a selected temperature. When the core is fully saturated, the permeability falls to a low value.

Next, let us consider the case where ac and dc components are present in a magnetic-core inductor. Figure 1-13 contains a family of curves that were

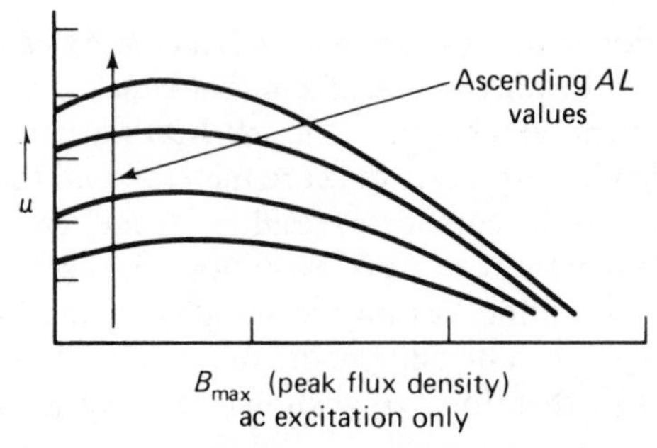

Figure 1-12 Curves demonstrating the relationship between permeability and flux density at peak ac excitation for several A_L factors. (Courtesy of Ferroxcube, Division of Amperex Corp.).

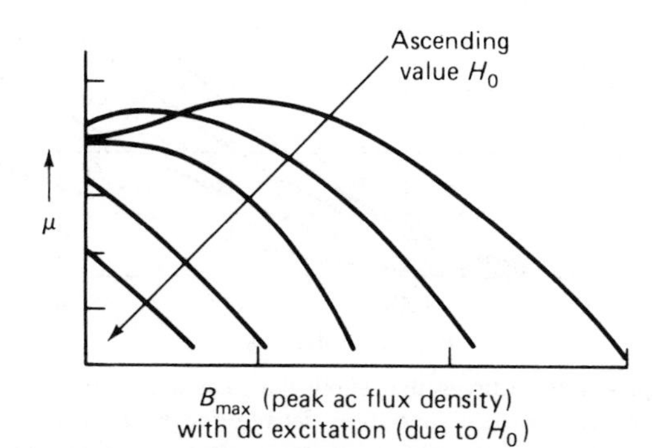

Figure 1-13 Curves showing the comparison between μ and B_{max} under various conditions of dc magnetization with simultaneous ac excitation. A single core material is represented here. (Courtesy of Ferroxcube, Division of Amperex, Corp.).

obtained from a specific type of core material with a defined shape and air gap. The curves compare μ to B_{max} for different values of dc magnetization of the core, which is subjected simultaneously to ac excitation. The dc magnetizing force is known symbolically as H_0 and uses oersteds as the reference units. H_0 is the unidirectional current flowing in an inductor or transformer winding. The curves illustrate clearly that dc magnetization lowers the μ_e of the core material, causing it to saturate at a lower value of peak flux density than would be the case under the conditions of Fig. 1-12. The flux density discussed here is a result of ac excitation only.

There is a relationship between core loss and flux density with respect to operating frequency. This consideration is shown by means of curves in Fig. 1-14. The curves were drawn for a specific core material, but the core structure and air-gap dimensions are not considered. It can be seen that the core loss is related directly to the flux density at selected frequencies. Curves of this variety can be established for core loss versus frequency at a number of flux densities. In this illustration the core loss is given in terms of mW/cm^2 of magnetic-core volume.

Finally, let us consider the matter of core losses versus temperature and flux density. Once again a particular type of core material is chosen for the collection of curve data, but the tests do not rely on the core shape or the air gap. Rather, the curves of Fig. 1-15 are based on core loss versus flux density at a number of core temperatures.

It was said earlier that part of the core heating can be related to the current flowing through the conductor, which is wound on the core. It is beneficial to select a wire gauge that will pass the ac and dc current with a minimum contribution to the self-heating characteristic. The dc resistance for an average winding can be determined by

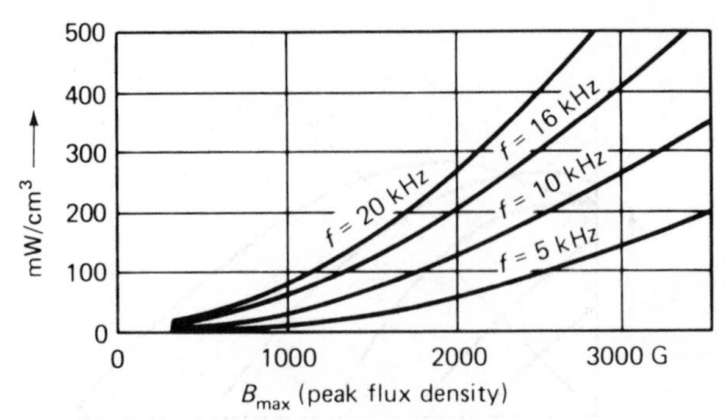

Figure 1-14 Curves that show the relationship among frequency, core loss, and flux density for a specified core material. The core loss is given in terms of mW/cm² of magnetic-core volume. (Courtesy of Ferroxcube, Division of Amperex Corp.).

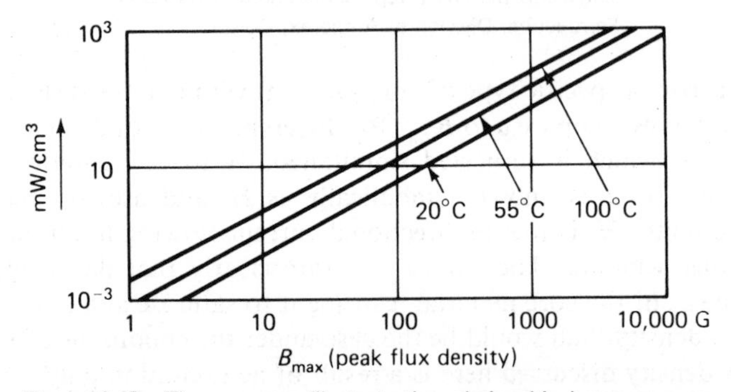

Figure 1-15 These curves illustrate the relationship between core loss and flux density at various temperatures. Again, a particular type of core material is the basis of the curves. (Courtesy of Ferroxcube, Division of Amperex Corp.).

$$R_{dc} = \frac{lwNr}{12{,}000} \text{ ohms}$$

where lw is the mean length of a single turn in inches, N the number of turns, and r the resistance of the wire in ohms per 1000 ft. Table 1-4 provides the appropriate data for finding the wire resistance, plus other information of interest concerning conductors which are used with magnetic-core materials.

1.4.2 Sample Calculation

Let us carry out a sample calculation for core selection at the input of an amplifier (Fig. 1-16). The operating frequency is 2.3 MHz and T1 will match the driver stage to the bases of two power transistors operating in Class C. T1 will function as a broadband conventional transformer.

TABLE 1-4 Data for Solid Copper Wire

AWG B & S Gauge	*DIAM (in.)*	*Cross Section (in.²)*	*Feet Per Ohm (20° C)*	*Ohms Per 1000 ft. (20°)*	*Amps For 1000A/in.²*	*Turns Per in.² H.F. Wire*	*Turns Per in.² S.F. Wire*
10	0.1019	0.00815	1001.	1.00	8.15	89	92
11	0.0907	0.00647	794.	1.26	6.47	112	118
12	0.0808	0.00513	630.	1.58	5.13	140	146
13	0.0719	0.00407	499.	2.00	4.07	176	180
14	0.0641	0.00322	396.	2.53	3.22	221	231
15	0.0571	0.00256	314.	3.18	2.56	259	275
16	0.0508	0.00203	249.	4.02	2.03	327	346
17	0.0453	0.00161	198.	5.06	1.61	407	432
18	0.0403	0.00127	157.	6.39	1.27	509	544
19	0.0359	0.00101	124.	8.05	1.01	634	679
20	0.0320	0.000804	98.5	10.7	0.804	794	854
21	0.0285	0.000638	78.1	12.8	0.638	989	1063
22	0.0254	0.000505	62.0	16.2	0.505	1238	1343
23	0.0226	0.000400	49.1	20.3	0.400	1532	1677
24	0.0201	0.000317	39.0	25.7	0.317	1893	2094
25	0.0179	0.000252	30.9	32.4	0.252	2351	2632
26	0.0159	0.000200	24.5	41.0	0.200	2932	3326
27	0.0142	0.000158	19.4	51.4	0.158	3711	4112
28	0.0126	0.000126	15.4	65.3	0.126	4581	5213
29	0.0113	0.000100	12.2	81.2	0.100	5621	6383
30	0.0100	0.0000785	9.69	104	.0785	7060	8145
31	0.0089	0.0000622	7.69	131	.0622	8455	10,097
32	0.0080	0.0000503	6.10	162	.0503	10,526	12,270
33	0.0071	0.0000396	4.83	206	.0396	13,148	15,615
34	0.0063	0.0000312	3.83	261	.0312	16,889	19,654
35	0.0056	0.0000248	3.04	331	.0248	21,163	25,531
36	0.0050	0.0000196	2.41	415	.0196	26,389	31,405
37	0.0045	0.0000159	1.91	512	.0159	31,405	39,570
38	0.0040	0.0000126	1.52	648	.0126	39,567	49,070
39	0.0035	0.00000962	1.20	847	.0096	53,855	65,790
40	0.0031	0.00000755	0.953	1080	.0075	65,790	82,180

The actual linear turns per square inch will vary with the manufacturer and the insulation thickness. Courtesy of Ferroxcube Corp.

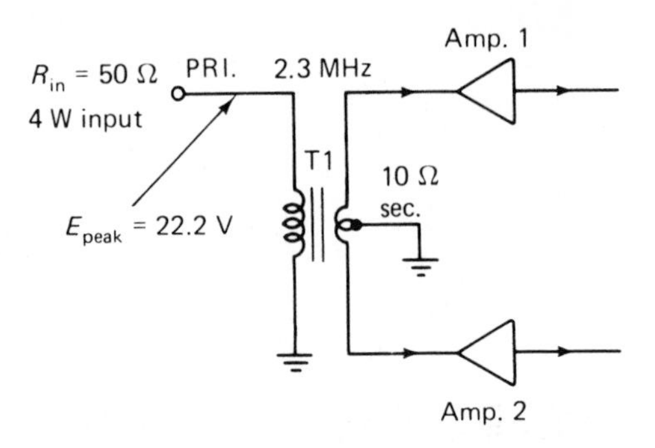

1. $X_{L\,(pri.)} \approx 4R_{in}$

   $\therefore\ X_{L\,(pri.)} = 200\ \Omega$

2. Select a core

   (Ferroxcube 76T188, type 4C4)

3. $L_{pri.} = \dfrac{X_L}{2\pi f} = \dfrac{200}{6.28 \times 2.3} = 13.8\ \mu H$

4. $\text{Turns} = 1000\sqrt{L\ \text{mH} \div A_L} = 1000\sqrt{0.0138 \div 70} = 14\ \text{turns}$

5. $B_{max} = \dfrac{E_{peak} \times 10^2}{4.44\ fnA_e} = \dfrac{22.2 \times 100}{4.44 \times 2.3 \times 14 \times 0.133} = 116.75\ \text{G}$

Figure 1-16 Sample core calculation for use at T1 while supplying excitation to a pair of Class C power amplifiers.

The primary reactance should be approximately four times the impedance of the driver—50 Ω. The required primary X_L is, therefore, 200 Ω, step 1 of Fig. 1-16. The next step calls for a core selection. A 0.5-in.-diameter Ferroxcube 76T188 is chosen arbitrarily for T1. Type 4C4 material should suffice, for it has a relatively high B_s (saturation density)—3000 G. In step 3 we determine the primary inductance required to satisfy the 200-Ω X_L at 2.3 MHz from step 1. This computes to 13.8 μH. The number of core turns is calculated in step 4, using the equation set forth earlier in the chapter. The core selected has an A_L index of 70; hence it will take 14 turns to provide an inductance of 13.8 μH.

To permit the B_{max} equation to be more workable for inductance in the μH range, it has been modified from the basic form to that of step 5 in Fig. 1-16. We are concerned with the *peak* ac voltage at T1, which at 4 W across a 50-Ω load is 22.2 V. (Although we have specified *peak* voltage in the example, rms voltage could be used. E_{peak} will allow for the worst-case condition.) The lowest operating frequency (f) is 2.3 MHz, the number of core turns (n) is 14, and the cross-sectional area of the chosen core in cm² is 0.133, as specified in the manufacturer's data sheet. B_{max} computes to 117 G. This is well below the rated B_s of 3000 G for the core, indicating the choice to be a good one. This will allow T1 to operate well within the linear region.

To illustrate the manner in which B_{max} is affected by changes in the values of the terms in the equation of Fig. 1-16, let us substitute 0.5 MHz for 2.3 MHz at f. The gauss becomes 537 rather than 117. Therefore, lowering the operating frequency while using a core that will yield proper X_L with the same number of turns (14) at 0.5 MHz will greatly elevate the B_{max}. Next, by using a core that requires 50 turns instead of 14 to obtain 13.8 μH, and changing nothing but n in the equation from Fig. 1-16, the B_{max} drops from 117 to 33.1.

Keeping all terms the same, but changing E_{peak} in Fig. 1-16 to 100, B_{max} increases to 526 G. Now, to demonstrate the effect of A_e, we will keep all terms the same but change A_e to 0.025, which represents a smaller toroid core. The B_{max} now rises to 629 G. Finally, to offer a more dramatic relationship of the terms in the equation to B_{max}, let us change E_{peak} to 100 and A_e to 0.025. This shows the effect of increased power versus decreased core cross-sectional area in cm². B_{max} under this set of conditions becomes 2798 G. The foregoing exercise demonstrates clearly that core choice is not a casual matter, particularly in circuits where power is present. The procedures for determining B_{max}, as given here, are applicable to all types of magnetic-core material, regardless of the shape chosen by the designer. It must be remembered that the foregoing discussion and the equation of Fig. 1-16 relate to B_{max} where only ac is present. The equation must be modified accordingly to accommodate the presence of ac and dc in the circuit. The basic equation for that condition was presented in Sec. 1.4.1.

1.4.3 Variations in the Equations

When the pertinent factors of a magnetic-core inductor are known, except for the value of L, this equation can be applied:

$$L = 0.4\pi N^2 \mu_e \left[\frac{A_e}{l_e}\right] \times 10^8 \quad \text{henries}$$

where L is the unknown inductance in henries, N the number of core turns, μ_e the effective permeability of the core, A_e the cross-sectional area of the core in cm², and l_e the effective magnetic length of the core in cm. The value of l_e can be found in the manufacturers' data sheets.

One can find B_{max} from known factors other than those given earlier in the basic equation. This variation is useful when specific numbers are known:

$$B_{max} = \left[\frac{2500E}{f\sqrt{LVe}}\right] \sqrt{\mu_e}$$

where f is in hertz, L in henries, V_e the effective core volume in cm^3 ($V_e = l_e \times A_e$), E the rms excitation voltage, and μ_e the effective permeability.

B_{max} can be defined also by

$$B_{max} = k_1 \sqrt{\mu_e}$$

$$\text{where } k_1 = \frac{2500E}{f\sqrt{LV_e}}.$$

It is useful also to know that $\mu_e = B_{max}^2/k_1^2$.

1.5 Volt-Ampere Ratings

A significant design parameter for a magnetic-core inductor which must handle power is the volt–ampere rating. This rating is of interest with respect to the lowest frequency at which the inductor will be utilized. Fundamentally, the volt–ampere rating for an inductor is determined by

$$EI = \frac{E^2}{2\pi fL} \quad \text{volt–amperes}$$

where E is in volts rms, I in amperes, f in hertz, and L in henries. It can be seen from this equation that the ratio of volt–amperes to frequency is

$$\frac{EI}{f} = \frac{E^2}{2\pi f^2 L} \quad \text{volt–amperes/Hz}$$

where the terms of the equation are in the same units as specified for the previous volt–ampere equation.

For the purpose of convenience, a new factor can be defined as k^2:

$$k_2 = \frac{2\pi EI}{f} = \frac{E^2}{Lf^2}$$

The maximum value for k_2, as applied to any core material, size, or shape can be expressed by

$$\frac{k_2}{2\pi} = \frac{15.6 \times 10^{-8}}{2\pi} \left[\frac{V_e (B_{max})^2}{\mu_e} \right] \quad \text{volt–amperes/Hz}$$

where B_{max} in this equation is the *maximum recommended* flux density for a specified core material, given by the manufacturer.

The reader may be wondering why the k_2 factor was brought into this discussion and what practical purpose it can serve. The k_2 information helps tell the designer which core size will fill the need for an inductor or transformer. Step 1 is to work the k_2 equation given previously. Once the value of k_2 is determined, the designer can select the smallest core that has a k_2 value that is equal to or greater than the operational characteristics required. The Ferroxcube literature lists the k_2 value and those of other factors used in the preceding equations of Sec. 1-5. Tables 1-5 through 1-8 contain this pertinent data for E cores, U cores, toroids, and pot cores.

1.5.1 A Sample k_2 Core Selection

Applying the k_2 factor is relatively simple if attention has been paid to the discussion in the foregoing section. A sample problem appears in Fig. 1-17 to demonstrate how a suitable toroid core can be chosen for a specified application. The diagram shows a 10-H inductor shunted across a 600-Ω audio line. The power level in the circuit is 0.5 W maximum. The lowest frequency of interest is 100 Hz.

Step 1 of Fig. 1-17 determines the rms voltage across L1. In step 2 the VA (volt–ampere) characteristic is calculated. From this we can learn the k_2 factor. Step 4 calls for inspection of the core characteristics in Table 1-7. Select a core k_2 that is equal to or greater than 0.00074. The nearest value (higher) is that of a K300502 core of the 3E type. The number of turns needed to obtain 10 H of inductance is provided by the A_L equation in step 5.

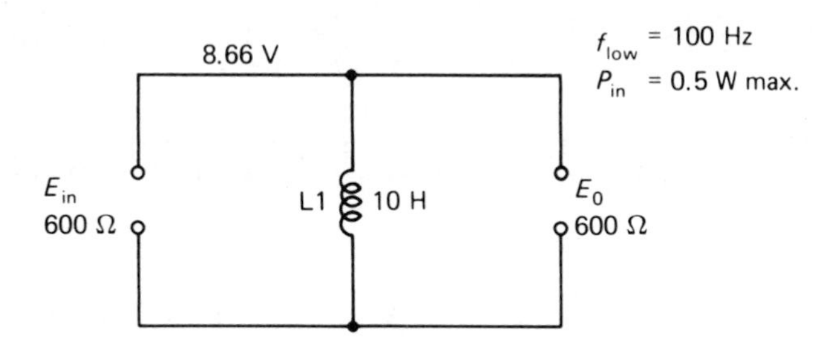

1. $E_{RMS} = \sqrt{WR} = \sqrt{0.5 \times 600} = 8.66$ V

2. $EI = \dfrac{E^2_{RMS}}{2\pi f_{Hz} L_H} = \dfrac{74.9}{6.28 \times 100 \times 10} = 0.0119$ VA

3. $k_2 = \dfrac{2\pi EI}{f} = \dfrac{6.28 \times 0.0119}{100} = 0.00074$

4. From Table 1.7 select core with $k_2 \geqslant 0.00074$. Nearest choice is a K300502, type 3E. Core $k_2 = 0.001$. $A_L = 1688$, $\mu_e = 2700$

5. $N = 1000\sqrt{L_{mH} \div A_L} = 1000\sqrt{10 \times 10^3 \div 1688} = 2434$

Figure 1-17 Mathematical progression for selecting a suitable core and wire turns for a typical audio reactor.

TABLE 1-5 E cores.

Core Part Number	Core Material Type	Con-figura-tion	e Ref.[a]	A_L (mH per 1000 turns)[a]	l_e (in. cm)	A_e (in.^{2} cm^2)	V_e (in.^{3} cm^3)	A_c (in.^{2} cm^2)	A_{CB} (in.^{2} cm^2)	$0.4A_{CB}$ (in.^{2} cm^2)	Approx. Data K_2[b]	Approx. Data P_0 (W)
814E250	3E2A	E-E	2800	2600	1.02	0.0298	0.0304	0.039	0.00810	0.0032	303×10^{-6}	0.32
					2.59	0.192	0.499	0.252	0.0523	0.0209		
		E-1	2200	2900	0.759	0.0314	0.0238	0.0195			301×10^{-6}	0.23
					1.93	0.202	0.390	0.126				
813E187	3E2A	E-E	3200	2300	1.54	0.0348	0.0538	0.082	0.0645	0.0258	468×10^{-6}	0.48
					3.91	0.224	0.882	0.529	0.416	0.166		
					3.91	0.224	0.882	0.529			407×10^{-6}	0.39
		E-1	2600	2600	1.10	0.0346	0.038	0.041				
					2.79	0.222	0.623	0.264	0.0320	0.0128	860×10^{-5}	0.65
813E343	3E2A	E-E	3200	4250	1.54	0.0639	0.0987	0.082	0.2060	0.0825		
					3.91	0.413	1.62	0.529			758×10^{-6}	0.48
		E-1	2600	4800	1.11	0.0639	0.0707	0.041				
					2.82	0.413	1.16	0.264	0.0470	0.0188	577×10^{-6}	0.60
873E189	3E	E-E	2080	1730	1.69	0.044	0.0745	0.0842	0.303	0.121		
					4.29	0.284	1.22	0.543			468×10^{-6}	0.43
		E-1	1850	2220	1.20	0.0452	0.0541	0.0421				
					3.05	0.291	0.887	0.272	0.0470	0.0188	741×10^{-6}	0.60
	3E2A	E-E	2800	2300	1.69	0.044	0.0745		0.303	0.121		
					4.29	0.284	1.22	0.543			569×10^{-6}	0.43
		E-1	2650	3200	1.20	0.0452	0.0541	0.0421				
					3.05	0.291	0.887	0.272	0.1034	0.0414	993×10^{-6}	0.79
812E250	3E2A	E-E	3320	3500	1.90	0.0621	0.118	0.125	0.667	0.267		
					4.83	0.400	1.94	0.806			832×10^{-6}	0.58
		E-1	2900	4100	1.0	0.0622	0.0867	0.0625				
					3.56	0.400	1.42	0.403	0.0472	0.0189	336×10^{-5}	1.21
206F440	3C5	E-E	(see note [c])		1.89	0.154	0.292	0.0992	0.304	0.122		
					4.80	0.993	4.79	0.638			261×10^{-5}	0.95
		E-1	(see note [c])		1.45	0.157	0.227	0.0496				
					3.68	1.01	3.72	0.319	0.0472	0.0189	246×10^{-5}	1.21
	3E	E-E	1900	4800	1.89	0.154	0.292	0.0992	0.304	0.1]2		
					4.80	0.993	4.79	0.638			207×10^{-5}	0.95
		E-1	1750	6100	1.45	0.157	0.227	0.0496				
					3.68	1.01	3.72	0.319	0.0472	0.0189	285×10^{-5}	1.21
	3E2A	E-E	2800						0.304	0.122		
					4.80	0.993	4.79	0.638			253×10^{-5}	0.95
		E-1	2500	8650	1.45	0.157	0.227	0.0496				
					3.68	1.01	3.72	0.319	0.152	0.608	185×10^{-5}	1.34

782E272	3E	E-E	2000	2180	2.60	0.0888	0.231	0.208	0.985	0.394		
					6.60	0.574	3.79	1.34			141×10^{-5}	0.94
		E-1	1900									
					4.65	0.588	2.74	0.670	0.152	0.608	215×10^{-5}	1.34
	3E2A	E-E	3000	3280	2.60	0.0888	0.231	0.208	0.985	0.394		
					6.60	0.574	3.79	1.34			166×10^{-5}	0.94
		E-1	2800	4450	1.83	0.0912	0.167	0.104				
					4.65	0.588	2.74	0.670				1.07
5690785	4A	E-E	≥ 450	≥1050	2.00	0.137	0.275	0.137				
					5.08	0.884	4.51	0.884				1.4
5690786	4B	E-E	≥ 195	≥ 452	2.03	0.141	0.288	0.141				
					5.16	0.909	4.72	0.909	0.323	0.129	408×10^{-5}	2.62
K540075	3E	E-E	≥1910	≥2575	3.98	0.163	0.651	0.429	2.08	0.833		
					10.11	1.05	10.67	2.77	0.289	0.115	628×10^{-5}	3.25
783E608	3E	E-E	2550	5900	3.83	0.281	1.08	0.429	1.86	0.756		
					9.73	1.81	17.71	2.77			537×10^{-5}	2.28
		E-1	2200	7500	2.64	0.280	0.739	0.214				
					6.71	1.81	12.12	1.38	0.289	0.115	885×10^{-5}	3.25
	3E2A	E-E	3400	8000	3.83	0.281	1.08	0.429	1.86	0.756		
					9.73	1.81	17.71	2.77			643×10^{-5}	2.28
		E-1	3200	10,800	2.64	0.280	0.739	0.214				
					6.71	1.81	12.12	1.38			872×10^{-5}	2.74
9F520	3C5	E-E	(see note [c])		3.02	0.251	0.757	0.300				
					7.67	1.62	12.41	1.94			852×10^{-5}	2.20
	3E	E-1	(see note [c])		2.61	0.283	0.740	0.150				
					6.63	1.83	12.14	0.968			605×10^{-5}	2.74
	3E	E-E	≥1500	≥4200	3.02	0.251	0.757	0.300				
					7.67	1.62	12.41	1.94			592×10^{-5}	2.20
		E-1	≥1500	≥5200	2.61	0.283	0.740	0.150				
					6.63	1.83	12.14	0.968			917×10^{-5}	4.57
5690718	3E	E-E	≥1970	≥3120	5.52	0.273	1.51	0.816				
					1402	1.76	24.76	5.26			593×10^{-5}	3.08
		E-1	≥1880	≥4500	3.54	0.263	0.930	0.408				
					8.99	1.70	15.25	2.63			162×10^{-4}	6.12
K540060	3E	E-E	2360	6000	14.68	2.66	39.19	5.73				

[a]Unless otherwise specified, μe and A_L values are nominal, with a +25% tolerance on A_L.

[b]For power-inductor designs, $k_2 = E^2/Lf^2$. For power-transformer designs, $K_2 = 2\pi E_p I_p/Mf$.

[c]For 3C5 material, the following applies: $\mu \geq 2100$ at 1000 G and 25°C; $\mu \geq 3000$ at 2000 G and 100°C. A nominal μ_e value of 2000 was employed in calculating the value of k_2 for 3C5 core material.

TABLE 1-6 U Cores.

APPROX. DATA

Core Part Number	Core Material Type	Con-figura-tion	μe Ref.[a]	A_L (mH per 1000 turns)[a]	l_e (in. cm)	A_e (in.^{2} cm^2)	V_e (in.^{3} cm^3)	A_C (in.^{2} cm^2)	A_{CB} (in.^{2} cm^2)	$0.4A_{CB}$ (in.^{2} cm^2)	k_2[b]	P_0 (W)
702U295	3E2A	U-U	1950	410	0.797	0.00525	0.00419	0.0265			602×10^7	0.09
					2.02	0.0338	0.069	0.171				
	3E2A	U-1	1650	370	0.613	0.00432	0.00265	0.0132	0.00740[b]	0.00296[b]	448×10^7	0.06
					1.56	0.0279	0.0435	0.0851	0.0477	0.0191		
376U250	3E2A	U-U	3600	2200	3.29	0.0625	0.205	0.375				
					8.37	0.400	3.36	2.418			159×10^5	1.21
	3E2A	U-1	3200	2500	2.53	0.0625	0.158	0.188				
					6.43	0.400	2.60	0.217			138×10^5	0.98
105F250	3E2A	U-U	2200	1100	1.16	0.0180	0.0209	0.0526				
					2.95	0.116	0.343	0.339			265×10^6	0.23
	3E2A	U-1	1950	1150	0.979	0.0180	0.0176	0.0263				
					2.49	0.116	0.289	0.1696			252×10^6	0.18
1F30	3C5	U-U	(see note[d])		4.31	0.134	0.579	0.562				
					10.9	0.864	9.50	3.624			667×10^5	2.32
	3C5	U-1	(see note[d])		3.33	0.137	0.456	0.281				
					8.46	0.884	7.48	1.812			525×10^5	1.74
1F31	3C5	U-U	(see note[d])		4.45	0.161	0.719	0.616				
					11.3	1.04	11.8	3.97			828×10^5	2.82
	3C5	U-1	(see note[d])		3.59	0.165	0.594	0.308				
					9.12	1.06	9.74	1.986			684×10^5	2.19
1F19	3C5	U-U	(see note[d])		7.13	0.227	1.62	1.950				
					18.1	1.46	26.6	12.577			187×10^4	5.49
	3C5	U-1	(see note[d])		5.73	0.231	1.32	0.975				
					14.6	1.49	21.6	6.288			152×10^4	4.08
1F10	3C5	U-U	(see note[d])		6.77	0.316	2.14	1.500				
					17.2	2.04	35.1	9.675			246×10^4	5.42
	3C5	U-1	(see note[d])		5.76	0.316	1.82	0.750				
					14.6	2.04	29.8	4.837			209×10^4	4.36
1F5	3C5	U-U	(see note[d])		12.4	1.00	12.1	5.000				
					31.5	6.45	198.	32.25			139×10^3	17.7
	3C5	U-1	(see note[d])		9.64	1.00	9.64	2.500				
					24.5	6.45	158.	16.125			111×10^3	13.45

[a]Unless otherwise specified, Ref. and A_L values are nominal with a +25% tolerance on A_L.

[b]For Power-Inductor designs, $k_2 = E^2/Lf^2$. For power-transformer designs, $k_2 = 2\pi E_p I_p/Mf$.

TABLE 1-7 Toroids.

Core Part Number	Core Material Type	μ_e Ref.	A_L (mH per 1000 turns) (+ 20%)	l_e (in. cm)	A_e (in.² cm²)	V_e (in.³ cm³)	A_C (in. cm²)	$0.4A_C$ (in.² cm²)	Approx. Data: k_2[a]	Approx. Data: P_0 (W)
213T050	4C4	125	24	0.422	0.00249	0.00105	0.00635	0.00254	b	b
				1.07	0.0161	0.0172	0.0410	0.0164	b	b
	3D3	725	140	1.07	0.0161	0.0172	0.0410	0.0164		
	3E2A	4470	850	1.07	0.0161	0.0172	0.0410	0.0164	653×10^{-8}	0.018
1041T060	4C4	125	25	0.532	0.00330	0.00176	0.0116	0.0463	b	b
				1.35	0.0213	0.288	0.0748	0.0298		
	3D3	725	144	1.35	0.0213	0.288	0.0748	0.0298	b	b
	3E2A	4495	890	1.35	0.0213	0.288	0.0748	0.0298	109×10^{-7}	0.025
266T125	4C4	125	55	0.852	0.0118	0.0100	0.0275	0.0110	b	b
				2.16	0.0760	0.164	0.177	0.0710		
	3D3	750	330	2.16	0.0760	0.164	0.177	0.0710	b	b
	3E2A	4830	2135	2.16	0.0760	0.164	0.177	0.0710	577×10^{-7}	0.05
768T188	3D3	750	415	1.19	0.0206	0.0247	0.0618	0.0247	b	b
				3.03	0.133	0.403	0.400	0.159		
	3E2A	5000	2750	3.03	0.133	0.403	0.400	0.159	137×10^{-6}	0.14
846T250	3E2A	5000	3055	2.18	0.0418	0.0908	0.228	0.0912		
				5.52	0.270	1.49	1.47	0.587	506×10^{-6}	0.32
K300502	3E	2700	1688	2.92	0.0582	0.169	0.440	0.176		
				7.43	0.375	2.78	2.84	1.14	101×10^{-5}	0.50
K300500	3E	2700	2422	3.58	0.101	0.360	0.643	0.257		
				9.11	0.650	5.92	4.15	1.660	214×10^{-5}	0.78
K300501	3E	2700	3639	3.58	0.151	0.542	0.643	0.257		
				9.11	0.975	8.88	4.15	1.660	321×10^{-5}	1.2
528T500	3C5			3.40	0.187	0.634	0.465	0.186		
				8.63	1.21	10.4	3.00	1.20	733×10^{-3}	1.3
400T750	3C5			5.00	0.281	1.40	1.27	0.508		
				12.7	1.81	23.0	8.20	3.28	162×10^{-4}	2.2
144T500	3C5			6.7	0.342	.30	1.84	0.736		
				17.1	2.21	37.8	11.86	4.744	265×10^{-4}	3.0

[a] For power-inductor designs, $k_2 = E^2/Lf^2$. For power-transformer designs, $k_2 = 2\pi E_p I_p / Mf$.

[b] Since this material is not recommended for high-flux density applications, no values are given for k_2 and P_0.

[c] For 3C5 material, the following applies: $\mu > 2100$ at 1000 G and 25°C; $\mu > 3000$ at 2000 G and 100°C.

Courtesy of Ferroxcube Corporation.

TABLE 1-8 Pot cores

Core Part Number	Core Material Type	Core Diam. mm	μe [a] Ref. [a]	A_L (mH per 1000 turns) [a] + 25%	l_e (in. cm)	A_e (in.^{2} cm^2)	V_e (in.^{3} cm^3)	A_C (in.^{2} cm^2)	A_{CB} (in.^{2} cm^2)	$0.4A_{CB}$ (in.^{2} cm^2)	APPROX DATA k_2 [b]	P_0 (W)
743P133-3E	3E	7	1100	576	0.408	0.0067	0.0027	0.0116	0.0056	0.0022	399×10^{-7}	0.57
					1.04	0.0433	0.045	0.0747	0.0361	0.0144		
332P133B4-3E	3E	9	1100	1350	0.559	0.0216	0.0120	0.0124	0.0057	0.0021	174×10^{-6}	0.86
					1.42	0.139	0.197	0.080	0.0368	0.0137		
1107P-L00-3B7	3B7	11	1430	1940	0.608	0.0259	0.0157	0.0164	0.0083	0.0033	378×10^{-6}	0.103
					1.54	0.167	0.257	0.105	0.0535	0.0214		
1107P-L00-3D3	3D3	11	635	865	1.54	0.167	0.257	0.105	0.0535	0.0214	c	c
1107P-L00-4C4	4C4	11	114	155	1.54	0.167	0.257	0.105	0.0535	0.0214	c	c
1408P-L00-3B7	3B7	14	1400	2240	0.771	0.0386	0.0297	0.0267	0.015	0.0060	665×10^{-6}	0.161
					1.96	0.247	0.487	0.172	0.0968	0.0387		
1408P-L00-3B9	3B9	14	1190	1910	1.96	0.247	0.487	0.172	0.0968	0.0387	c	c
1408P-L00-3D3	3D3	14	705	1130	1.96	0.247	0.487	0.172	0.0968	0.0387	c	c
1408P-L00-3E	3E	14	1300	2080	1.96	0.247	0.487	0.172	0.0968	0.0387	365×10^{-6}	0.161
1408P-L00-4C4	4C4	14	125	200	1.96	0.247	0.487	0.172	0.0968	0.0387	c	c
1811P-L00-3B7	3B7	18	1740	3680	1.02	0.0672	0.0683	0.0458	0.029	0.0116	123×10^{-5}	0.265
					2.59	0.433	1.12	0.296	0.187	0.0748		
1811P-L00-3B9	3B9	18	1250	2630	2.59	0.433	1.12	0.296	0.187	0.0748	c	c
1811P-L00-3D3	3D3	18	735	1550	2.59	0.433	1.12	0.296	0.187	0.0748	c	c
1811P-L00-3E	3E	18	1660	3500	2.59	0.433	1.12	0.296	0.187	0.0748	658×10^{-6}	0.265
1811P-L00-4C4	4C4	18	125	265	2.59	0.433	1.12	0.296	0.187	0.0748	c	c
2213P-L00-3B7	3B7	22	1825	4650	1.23	0.0985	0.122	0.0652	0.046	0.0184	208×10^{-5}	0.392
					3.12	0.635	1.99	0.420	0.297	0.118		
2213P-L00-3B9	3B9	22	1275	3250	3.12	0.635	1.99	0.420	0.297	0.118	c	c

2213P-L00-3D3	3D3	22	705	1800	3.12	0.635	1.99	0.420	0.297	0.118	c	c
2213P-L00-3E	3E	22	1725	4400	3.12	0.635	1.99	0.420	0.297	0.118	112×10^{-5}	0.392
2213P-L00-4C4	4C4	22	120	300	3.12	0.635	1.99	0.420	0.297	0.118	c	c
2616P-L00-3B7	3B7	26	1880	6000	1.47	0.147	0.217	0.0890	0.063	0.025	362×10^{-5}	0.555
					3.73	0.948	3.56	0.574	0.406	0.163		
2616P-L00-3B9	3B9	26	1380	4390	3.73	0.948	3.56	0.574	0.406	0.163	c	c
2616P-L00-3D3	3D3	26	735	2340	3.73	0.948	3.56	0.574	0.406	0.163	c	c
2616P-L00-3E	3E	26	1830	5850	3.73	0.948	3.56	0.574	0.406	0.163	190×10^{-5}	0.555
2616P-L00-4C4	4C4	26	120	390	3.73	0.948	3.56	0.574	0.406	0.163	c	c
3019P-L00-3B7	3B7	30	2020	7580	1.77	0.214	0.380	0.124	0.091	0.0364	608×10^{-5}	0.767
					4.50	1.38	6.23	0.800	0.587	0.235		
3019P-L00-3B9	3B9	30	1480	5750	4.50	1.38	6.23	0.800	0.587	0.235	c	c
3019P-L00-3D3	3D3	30	745	2820	4.50	1.38	6.23	0.800	0.587	0.235	c	c
3622P-L00-3B7	3B7	36	2000	9660	2.08	0.313	.651	0.167	0.116	0.0464	102×10^{-4}	1.06
					5.28	2.02	18.7	1.08	0.748	0.299		
3622P-L00-3B9	3B9	36	1440	7050	5.28	2.02	10.7	1.08	0.748	0.299	c	c
3622P-L00-3D3	3D3	36	745	3580	5.28	2.02	10.7	1.08	0.748	0.299	c	c
3622P-L00-3E	3E	36	2250	10,800	5.28	2.02	10.7	1.08	0.748	0.299	464×10^{-5}	1.06
4229P-L00-3B7	3B7	42	2100	10,300	2.68	0.413	1.11	0.301	0.217	0.0868	165×10^{-4}	1.60
					6.81	2.66	18.10	1.94	1.40	0.560		
K5 350 56-3E	3E	45	2400	13,500	2.56	0.451	1.15	0.251	0.160	0.0640	764×10^{-5}	1.59
					6.50	2.91	18.8	1.62	1.03	0.412		
K5 350 11-3E	3E	66	2440	18,200	4.85	1.11	5.38	.857	0.620	0.248	352×10^{-4}	4.42
					12.30	7.16	88.2	5.53	4.00	1.60		

[a]Per pair of cores.

[b]For power-inductor designs, $k_2 = E^2/Lf^2$. For power-transformer designs, $K_2 = 2\pi E_p I_p/Mf$.

[c]Since this material is not recommended for high-flux density applications, no values are given for k_2 and P_0.

Courtesy of Ferroxcube Corporation.

The design objective has been satisfied when 2434 turns of wire are layer-wound on the toroid core. The progression of design is patterned after information contained in Ferroxcube *Bulletin 330-A*. In a practical situation the core used should be physically larger than that chosen in Fig. 1-17, even if the k_2 rating is substantially higher. This will facilitate easier winding and permit the use of larger wire gauges than would be required for the core specified.

1.5.2 Power Rating

Tables 1-5 through 1-8 contain a set of P_0 factors (power dissipation factor) in watts. This is the power-dissipation level which will cause a temperature rise of 50 °C above the ambient level within a specified core. P_t (total power dissipation) is a significant factor also. It is the total power dissipated in an inductor, constituting winding loss and core loss. The dielectric losses are not part of the P_t factor.

Once the wire size and number of coil turns have been chosen, the designer can find the resistance of the winding (Table 1-4) and estimate the total winding copper loss. The core-loss factor can be obtained from the manufacturer's published curves for the material used. These data, plus the winding-loss data, provide the information needed to calculate the P_t.

Once the power is known, the temperature-rise trait can be taken from

$$T_{\text{rise}} = \frac{50P_t}{P_0}$$

where the factor P_0 is given for each core type in the tables. Once this information is acquired, the designer can make final adjustments (empirical) to compensate for the temperature-rise effects on μ_e and the inductor Q. It may at this juncture be necessary to modify the number of coil turns slightly to meet the initial design objective. Typically, assuming that the design was done carefully at the beginning, the changes caused by heating should be of low percentage.

1.6 Inductance versus AC/DC Excitation

It is a common design characteristic to have a large component of dc excitation present along with the ac excitation related to a given power inductor. Figure 1-18 illustrates a typical application, where V_{cc} is supplied to an RF power transistor through a magnetic-core choke inductor, L1. The same condition does not prevail with respect to L2 and L3 of Q1 because C1 serves as a blocking capacitor. L2 and L3 are subjected to ac excitation only.

It is not uncommon for the dc excitation component to exert considerably more magnetomotive force than is applied by the ac-derived ex-

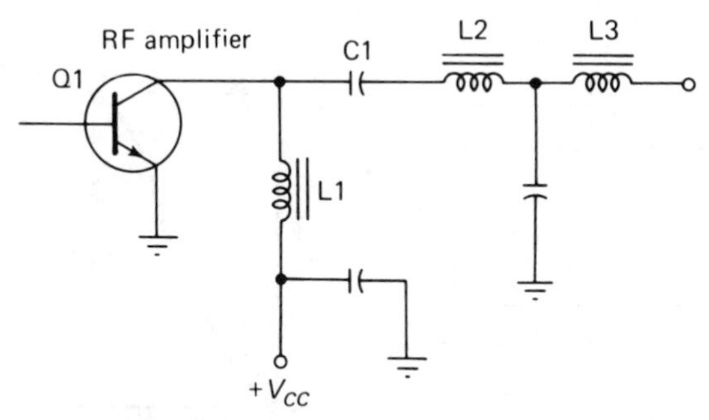

Figure 1-18 Example of a magnetic-core inductor used to supply dc operating voltage to the collector of an RF-amplifier transistor. Ac and dc components are present in L1, but only ac is a consideration at L2 and L3.

citation. In a design of this variety the attendant problems can be minimized to a considerable extent by introducing an air gap in the magnetic circuit. The advantage is that a smaller core can be used than would be possible with a closed magnetic circuit, such as a classic toroid core. The principle of air-gap inclusion is shown in Fig. 1-19a.

The designer must be aware that the introduction of an air gap has a marked effect on the inductance that would be present in a closed-core circuit. The air gap creates what can be termed a "hybrid flux path." This is because part of the flux path is via a fairly long ferrite circuit of high μ, while another part of the path is through a fairly short low-μ air gap. The effective length of the magnetic path under these conditions can be expressed as

$$l_e = l_m + \mu\, l_g$$

where μ is the permeability of the material, l_m the length of the magnetic path (normally specified as l_e for zero gap), and l_g the length of the core gap. All lengths are in centimeters. The air-gap μ is considered as 1, or unity.

1.6.1 Employment of the Hanna Curve

Air gaps in cores are used with regularity in the design of power chokes, filter inductors, and transformers. We have just outlined the fundamentals of designing with air gaps, but finding the exact air-gap dimension for a given optimum condition versus core material, inductance, and dc current is a tedious undertaking at best. A simplified method was disclosed by C.R. Hanna in his 1927 AIEE Winter Convention paper. His method enables the designer to derive the best air gap for a specific group of requirements. From his technique, Fair-Rite Products Corp. constructed a Hanna curve for use with its Fair-Rite No. 77 core material, a popular type for use in power inductor and transformer applications.

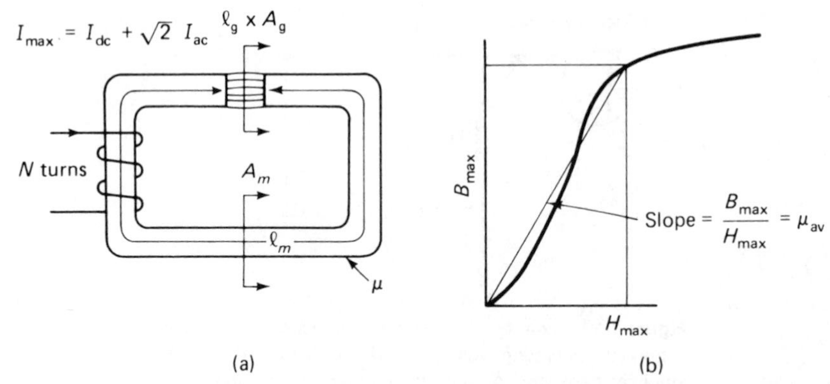

Figure 1-19 The concept of a magnetic core with air gap is shown in (a). The curve in (b) illustrates how μ_{av} departs from a straight line. However, the straight line is used as the basis for μ_{av} in a practical situation.

During our discussion earlier, we established that dc current flowing through the winding of a ferromagnetic inductor would saturate the core to some extent, thereby reducing the inductance. This effect is demonstrated clearly by the curve marked "Air gap = 0" in Fig. 1-20. Not only does an air gap lower the inductance, it decreases the inductance more and more as the gap is made wider. But the benefit realized from increasing the width of the gap is greater direct-current capability before core saturation takes place. The various air-gap curves in Fig. 1-20 illustrate this principle rather clearly.

A typical method for finding the optimum core gap of a given inductor is that of trial and error—a truly empirical approach. The usual approach is to chose a core that will yield the required amount of inductance, based on a wire gauge that will handle the dc current without excessive heating or IR voltage drop. Next, the assembled inductor is subjected to the normal level of dc current to learn if saturation is occurring. If saturation is noted, assorted air gaps are tried experimentally until one dimension prevents saturation. The shortcoming is a need to experiment further—adding turns on the coil to compensate for the lost inductance. This can require a larger core in some instances. The foregoing routine is repeated until the proper core gap and core size is found. This means a core that will not saturate when dc bias is applied. The process *is* a tedious one, indeed!

The procedure is simplified by using the curves of Fig. 1-20 for the core that is chosen first. The designer can pick a curve that is flat almost to the maximum amount of dc bias current in the coil. The curve chosen will show the available inductance per coil turn versus the air gap selected. From this information it is a simple task to calculate the total turns required for the target inductance. If, of course, the specified number of turns will not fit on the core, a larger one must be chosen. This will necessitate a repeat of the selection procedure and the use of a new set of curves.

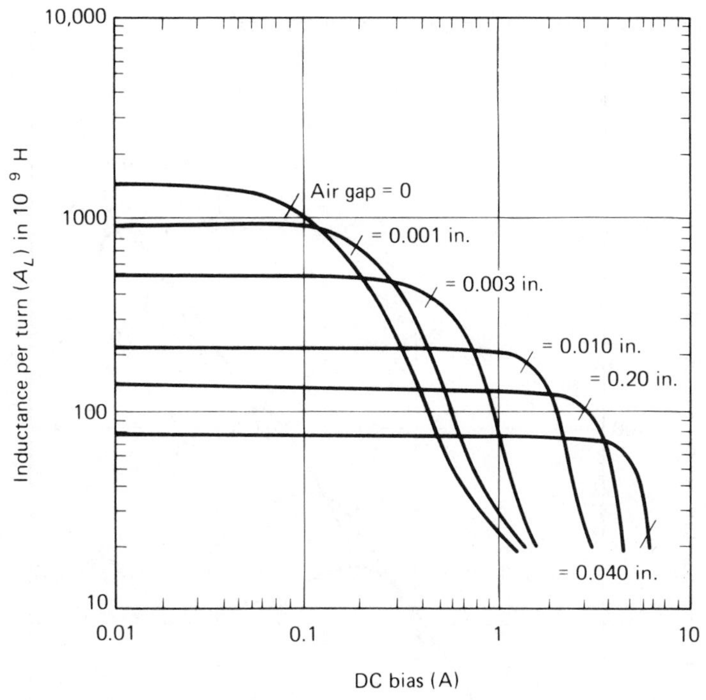

Figure 1-20 Curves that show the effect of increasing air gap versus inductance. (Courtesy of Fair-Rite Products Corp.).

Air-gap selection by means of the Hanna curve eliminates the need for a separate set of curves for each core size, with respect to dc bias and inductance. But, of greater significance, it negates the need for trial-and-error determinations.

1.6.2 Constructing a Hanna Curve

Observation of the curves in Fig. 1-21 illustrates the construction of the Hanna curve. Note the similarity between these curves and those of Fig. 1-20. The vertical scale is different, however, as a result of multiplying it by the dc current squared. The dashed line of Fig. 1-21 provides a point of tangency on each of the curves which indicates the optimum inductance for the given air gap and dc bias. The dashed line is shown with slight alterations in Fig. 1-22. In this presentation it is the classic Hanna curve.

In Fig. 1-22 the scales are normalized as a result of dividing the vertical scale by the effective core volume. The horizontal has been changed to indicate the magnetizing field in oersteds. The curve is characteristic of the Fair-Rite No. 77 core material. It can be used for any core size in that grade of material. The air-gap tangency points indicated in Fig. 1-23 are identified in Fig. 1-22 by means of the short marker lines which are perpendicular to

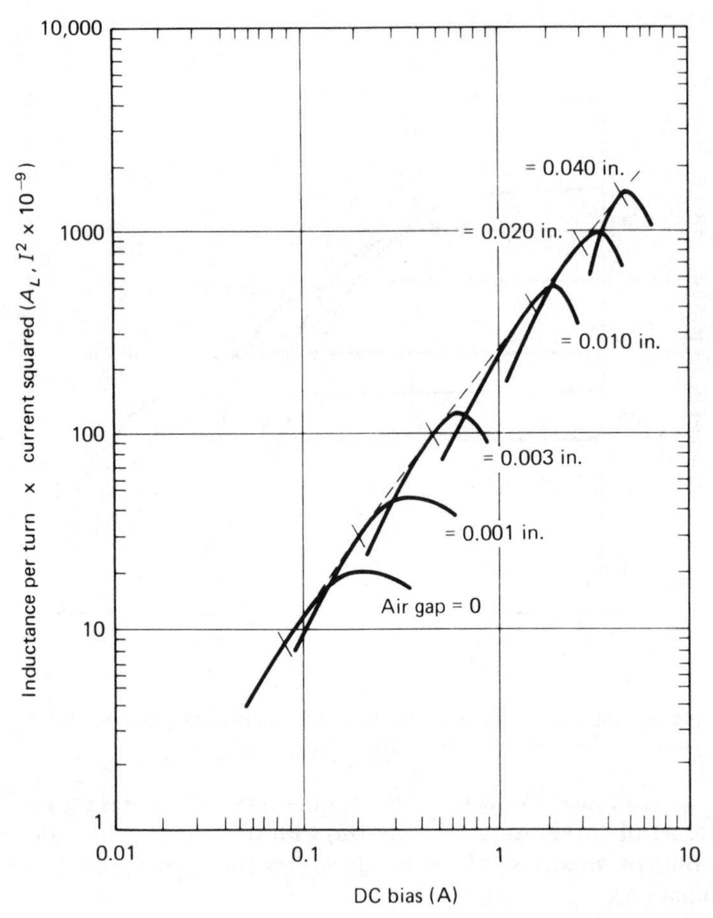

Figure 1-21 This group of air gaps form a Hanna curve. The curve is strikingly similar to that of Fig. 1-20. (Courtesy of Fair-Rite Products Corp.).

the curve. These points are based on the value of the air gap, divided by the effective path length of the core (l_e) in centimeters.

Normally, the first step in using the Hanna curve dictates selecting a core that will yield the required amount of inductance versus an acceptable wire gauge. Following this it is possible to calculate the vertical factor of the curve by

$$V_f = \frac{L \times I^2}{V_e}$$

where L is the inductance in henries, I the dc current in amperes, V_e the effective core volume in cm³, and V_e the vertical factor.

Once the vertical factor is derived, a horizontal line is drawn from the vertical axis until it meets the Hanna curve. At this point of intersection we

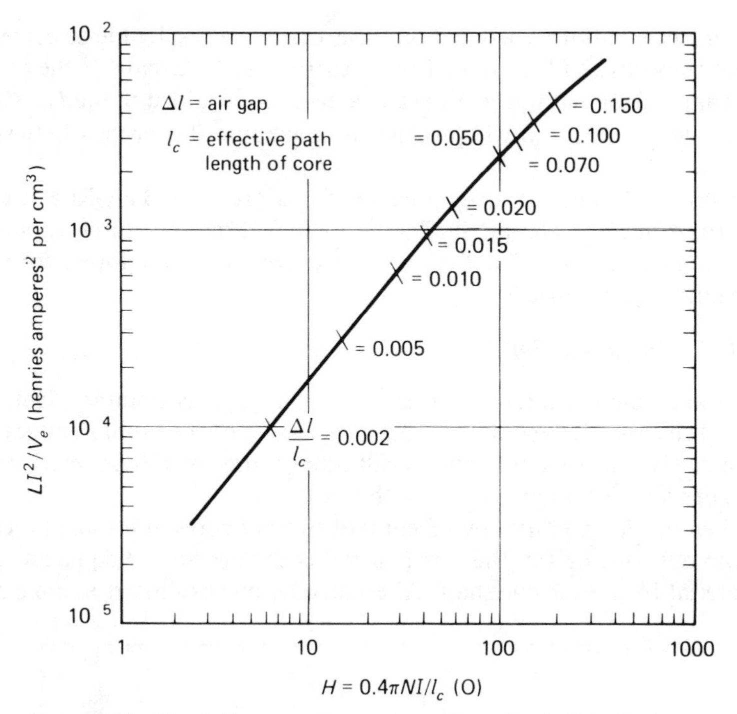

Figure 1-22 Classic Hanna curve as extracted from the significant points on the curve in Fig. 1-21. (Courtesy of Fair-Rite Products Corp.).

have the value of the magnetizing field in oersteds (produced in the core). The equation for this is

$$H = \frac{0.4 \times N \times I}{l_e}$$

where H is in oersteds, L the inductance in henries, I the dc current in amperes, and l_e the effective path length of the core in centimeters.

Once this value is obtained, it is possible to calculate the number of turns required. As mentioned earlier, if the turns will not fit on the chosen core, a larger one must be selected and the process repeated.

The air-gap factor is found at the intersection point of the horizontal line and the Hanna curve. The air-gap factor can be determined by

$$\mathrm{AG}_f = \frac{\Delta l}{l_e}$$

where AG_f is the air-gap factor, Δl the air gap in inches, and l_e the effective path length of the core in inches.

The air-gap factor varies in a nonlinear manner. For this reason it is difficult to use interpolation between the defined points on the curve. The foregoing calculations are based on the premise that negligible ac current is present respective to the amount of dc current in the circuit. If the ac current is appreciable, however, its peak value must be added to the I_{dc}. This is because the *total current* will produce the maximum flux density in the core material.

Figure 1-23 contains air-gap curves for a group of Fair-Rite E cores made from No. 77 material. Similar curves for the manufacturer's pot cores are presented in Fig. 1-24. Both sets of curves were developed from the Hanna curve procedure.

1.6.3 Actual Air Gap

We are now concerned with the *total* air gap, as obtained from the curves. It can be effected by grinding away a part of the E-core center post. Alternatively, most manufacturers will supply cores with the prescribed air gap. The same technique is used with pot cores.

When the designer uses two E cores or two pot cores, it is more practical to grind only one-half of the core pair rather than removing equal amounts of material from each core half. Alternatively, one can insert nonmagnetic

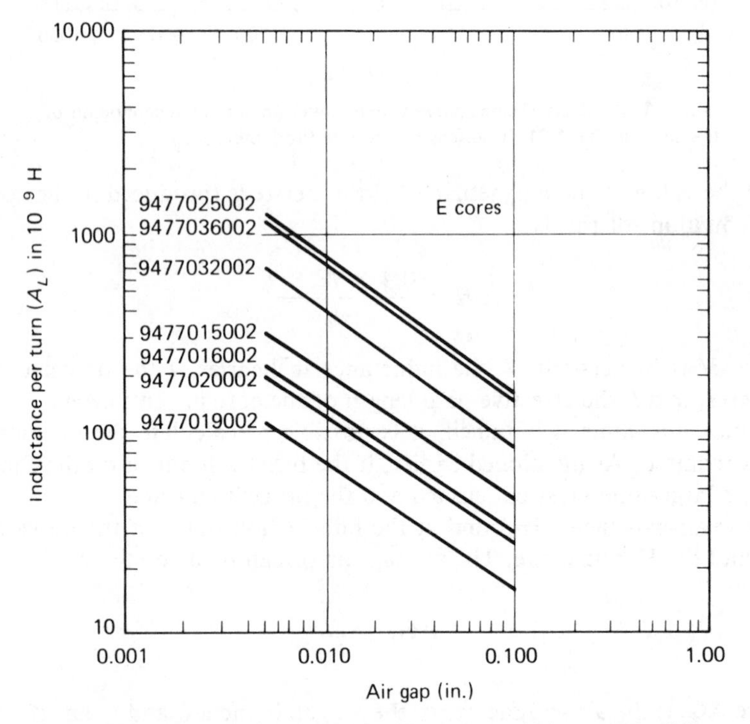

Figure 1-23 Curves for a group of Fair-Rite E cores which are made of No. 77 (1800 μ_i) ferrite material. (Courtesy of Fair-Rite Products Corp.).

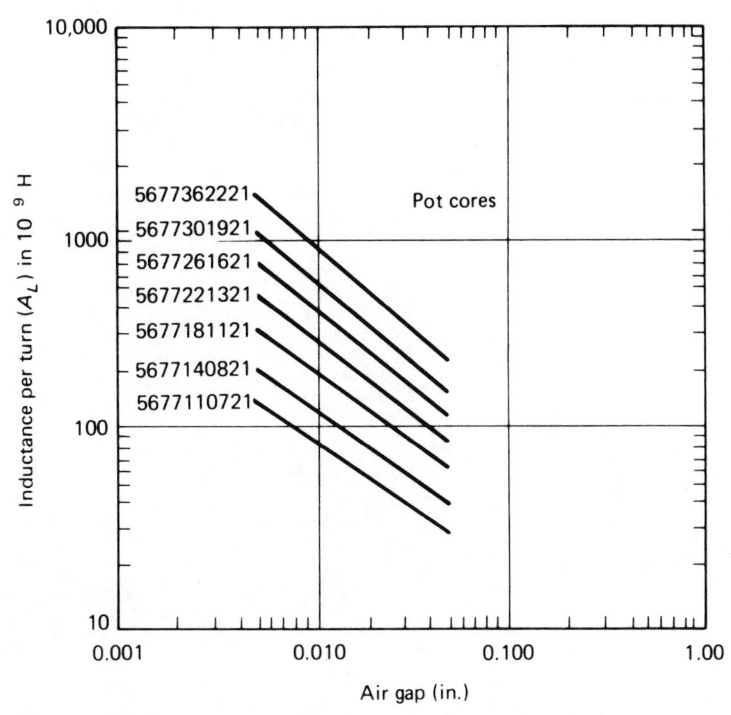

Figure 1-24 Fair-Rite Products Corp. curves for No. 77 (1800μ_i) pot cores. (Courtesy of Fair-Rite Products Corp.).

shim stock between the core halves to provide an air gap. In such cases the stock is inserted between the outer posts of the E cores and between the halves of the pot cores along the outer perimeter of the cores. When this technique is applied, it is necessary to use shims which are half as thick as the desired air gap. This is because the center post of the core is elevated by the same amount as the outer legs or walls, creating an effective gap which is twice the thickness of the shim stock.

2

APPLICATION OF RODS, BARS, AND SLUGS

We discussed in Chapter 1 the basic principles and design considerations for the use of magnetic cores. Those details will not be repeated except when it is useful to demonstrate a design procedure. In this chapter we address the practical applications in which one might use magnetic-core materials in the form of rods, bars, and slugs. For the most part, circuits that employ these physical forms are in the narrow-band category, principally for use in the radio-frequency spectrum. For the purpose of design simplification we will work with approximations and rules of thumb where it is practical, sidestepping the more rigorous mathematical solutions. Fortunately, the performance results obtained from circuits that are based on well-founded rules of thumb are entirely satisfactory for most prototype work. Refinement of the basic design can always be accomplished, if needed, just prior to the preproduction stage, or the "proof-build" stage, as it is sometimes called.

2.1 Rods and Bars

Ferrite rods and bars are available in standard lengths up to 8½ in. The rods can be obtained with diameters from ¼ in. to ¾ in. Bar stock is available in thicknesses from ⅛ to ¼ in., with widths from ⅜ to ⅞ in. The bar stock is sold in the flat or rectangular format (Fair-Rite Classes 34 and 35). Rod materials can be purchased in the solid or hollow formats (Fair-Rite Classes 30 and 31, respectively).

Ferrite rods and bars are used principally as magnetic cores for radio antennas from the VLF spectrum well into the VHF range. The advantages of these physically small antennas are high Q, compactness of assembly, and portability of equipment. The rods and bars manufactured by Fair-Rite Products Corp. are in materials 61 (μ_i = 125, 2350 G) and 64 (μ_i = 250, 2200 G). Specific rod and bar sizes are available in materials 33 (μ_i = 800), 63 (μ_i = 40), and 68 (μ_i = 20). Number 63 stock is optimum for use at approximately 30 MHz and the 68 material is best suited for applications at roughly 150 MHz. Figure 2-1 shows various types of rod and bar material in photographic form.

2.1.1 Characteristics of Rods and Bars

Rod material is available in small dimensions for use as core material in slug-tuned inductors. Although the general rules for application of slugs and rods are the same, this section will treat the larger rods and bars. Slug-tuned inductors and transformers will be discussed in detail later in this volume.

It is important that we keep in mind that the μ_i rating is that of the material from which a core is made. The size and shape of the finished product dictates the more significant parameter, μ_e, or *effective permeability*. The factor of interest in establishing the μ_e of a specified core μ is the *diameter-to-length ratio* of the rod or bar. Figure 2-2 contains a family of curves that illustrates this concept. The curves have been drawn for a standard group of μ_i factors for ferrite materials. The vertical scale displays the μ_e for each kind of material versus the rod length/rod diameter *(l/d)*. It becomes apparent when examining the curves that the rods or bars with high initial permeabilities (40 and greater) decrease the most in μ_e when the length is short. It can be seen in Fig. 2-2 that a 0.25-in.-diameter rod, 8½ in. long, would be needed to obtain an effective permeability of 30, even though the rod material being used exhibited an initial permeability of 40.

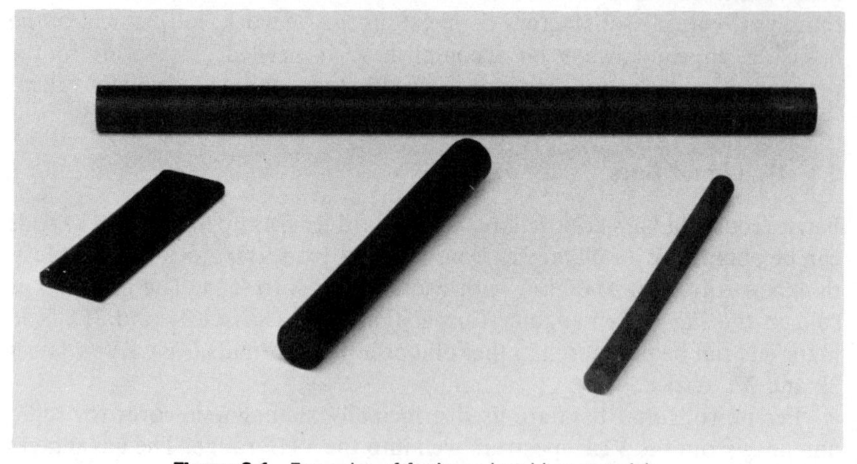

Figure 2-1 Examples of ferrite rod and bar materials.

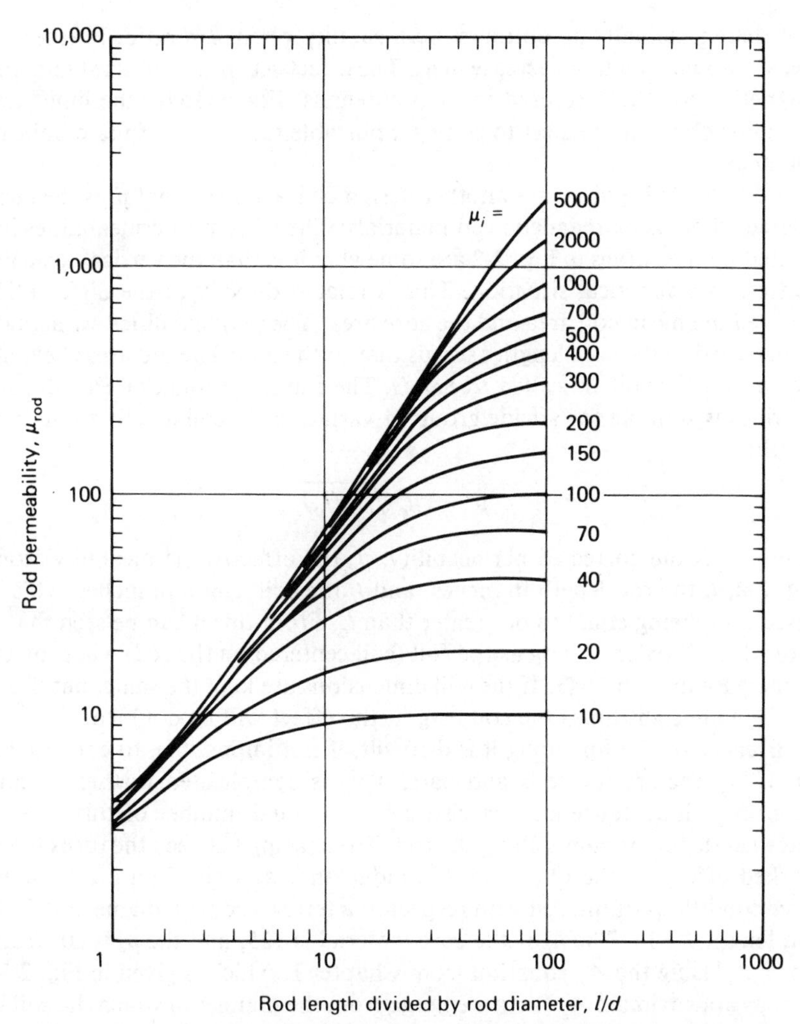

Figure 2-2 Family of curves that illustrate the relationship between rod length and diameter to permeability. (Courtesy of Fair-Rite Products Corp.).

On the other hand, if a lower μ_i is selected (10) and the form factor is reduced to 15 (0.25-in. in diameter, 3.75 in. in length), the μ_e remains the same as the μ_i. At the upper end of the initial-permeability range it becomes a practical impossibility to realize a rod whose μ_e is comparable to the μ_i of its material. For example, a rod that measured 0.25 × 25 in. (form factor = 100), and whose μ_i is 5000, would exhibit a μ_e of only 1800. Because of this phenomenon, the designer will find the curves in Fig. 2-2 of considerable value.

If we were to extend the curves of Fig. 2-2 to the right, it could be seen that there is actually an optimum permeability versus l/d ratio. Further increases would lead to a decrease in μ_e. The effect sets practical size limits for magnetic rods that are used in loop antennas. Fortunately, the limits are workable ones with respect to compact portable radios that contain built-in antennas.

Polydoroff [1] discusses another design characteristic that must be considered when using magnetic-rod materials: The effective permeabilities indicated by the curves in Fig. 2-2 are somewhat less than the vertical scale indicates, in a practical situation. This is related directly to the differential between the mean coil area and the core area. The permeabilities are actually increased as the core length exceeds that of the coil. The rod areas beyond the ends of the coil are called *free ends*. The empirical equation Polydoroff provides is suitable for a wide group of variations in coil length versus rod length:

$$\mu' = \mu_e \sqrt[3]{(l_r/l_o)}$$

where μ' is the corrected permeability, μ_e the effective permeability from Fig. 2-2, l_r the rod length in inches, and l_c the coil length in inches. This is based on l_r being equal to or greater than l_c. From this it can be seen that a core which is twice as long as the coil (coil centered on the rod) will provide an increase in μ_e of 26%. If the coil dimensions are kept the same, but if the rod is made eight times the coil length, the effect will be doubled.

Because of the foregoing it is difficult, if not impossible, to construct a set of A_L factors for rods and bars. This is complicated further by differences in inductance characteristics for an equal number of turns being placed at different points along the rod. The spacing between the turns has a marked effect on the Q and on the inductance as well. Figure 2-3 shows three conditions of interest with respect to a ferrite rod with diameter 0.5 in. and length 7.5 in. The μ_i of the core material is 125, and the μ_e is 60, from Fig. 2-2. Using the A_L equation from Chapter 1, which is given in Fig. 2-3, various approximate A_L factors emerge from the manner in which the coil is placed on the rod. These approximations hold for a rod of the stated dimensions and μ_i. Rods of different dimensional format would result in radically different A_L numbers. The measurements were made by the author while using a Hewlett-Packard 4342A Q meter. The rod material for these tests was an Amidon Associates R61-500X7.

2.1.2 Q and L versus Coil Placement

There are some schools of thought which suggest that a coil placed at one extreme of a selected rod will yield greater Q than when the coil, pruned for the same inductance, will exhibit at the physical center of the rod (Fig. 2-3a and b). The graphical exhibit of Fig. 2-5 shows that this is a false assumption. While using the same rod that served in the tests of Fig. 2-3, an

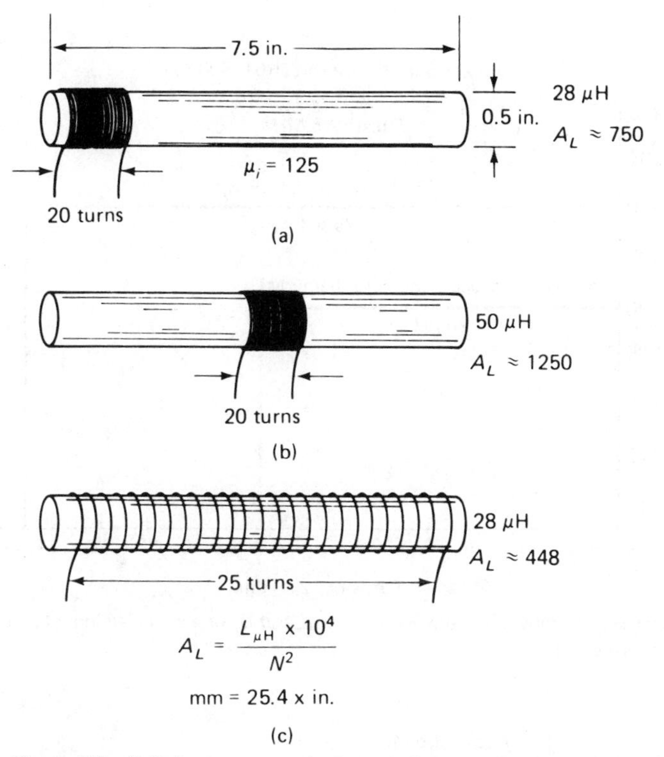

Figure 2-3 Coil placement on a ferrite rod of given μ_i has a direct bearing on the A_L factor.

investigation of the unloaded Q was carried out. Figure 2-5 shows a dramatic difference in Q_u between end placement and center positioning of the coil. With the coil set to provide 27 μH of inductance at the end of the rod, the Q is 80 for 20 turns of No. 20 Formvar wire, close-wound tightly on the core. The coil was trimmed to yield 27 μH at the center of the rod, and the Q rose to 190 for 15 turns.

Another experiment was conducted with the Q meter. The results are seen in the graphic illustration of Fig. 2-4. A 28-μH inductance was obtained by spreading 25 turns of No. 22 Teflon-insulated hookup wire across the entire length of the ferrite rod, as in Fig. 2-3c. The effect shown in Fig. 2-5 was reversed. A Q_u of 240 was obtained under the spread condition, whereas the Q_u declined only slightly to a value of 230 when 15 turns of the same wire were placed at the center of the rod in close-wound fashion. In this example the copper conductor was spaced by the thickness of the Teflon insulation. This accounts for the higher Q reading for the center-positioned coil respective to the one of Fig. 2-5, which used thin insulation (Formvar). It can be seen from this that the Q dropped only 4% when the coil was moved to the center. The Q declined 58%, however, in the tests of

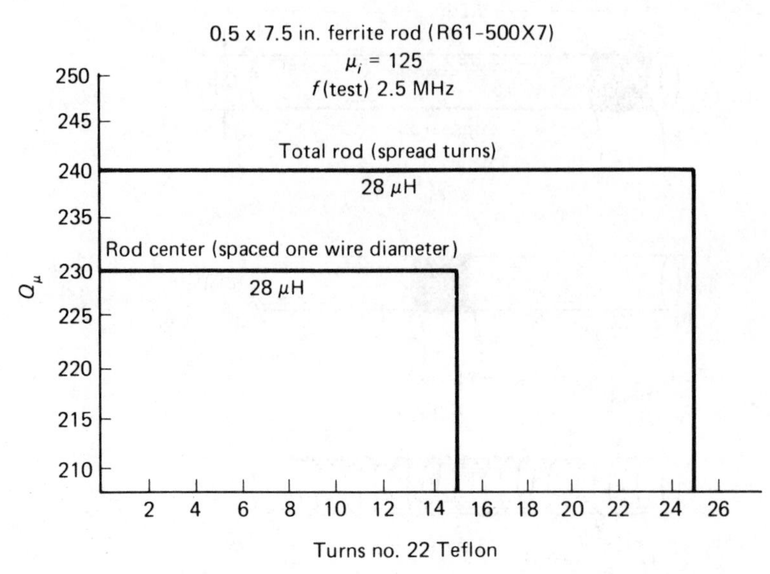

Figure 2-4 Comparison between coil turns and Q for a specified inductance on a given ferrite rod.

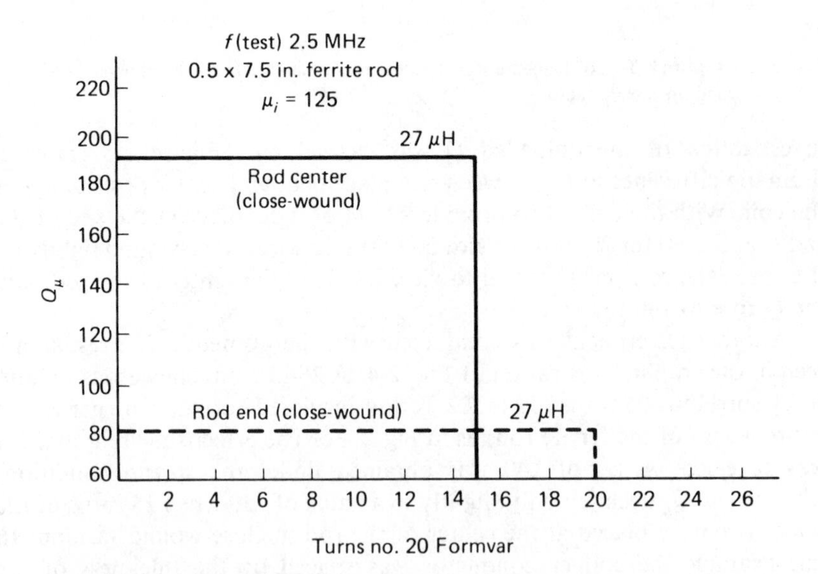

Figure 2-5 The Q_u of a 27-uH inductor is 80 when the winding is placed at the end of a ferrite rod. A Q_u of 190 was obtained by placing the 27-uH winding at the center of the rod.

Fig. 2-5, where the actual conductor turns were closer together and insulated from one another by a material with a lower dielectric constant.

Further investigation was conducted to compare the placements of coils with identical turns and dimensions, versus end and center location. First, a one-turn coil was moved from the rod center to the extreme end of the core. Figure 2-6 displays the results. In this case the Q_u was slightly higher at the end (155) as opposed to the center (145), a variation of only 6%. The inductance at the center was 8.4% greater than at the end—1.19 and 1.09 μH, respectively. This demonstrates that there is a definite relationship between μ_i and the number of turns when attempting to find a workable A_L, along with the factors discussed earlier. If one were to pick an A_L value referenced to a single turn at the center of the rod, a value of 11,900 would result, based upon $A_L = L_{\mu H} N^2$. The difference between this and the value arrived at when using 20 turns (Fig. 2-3b) is startling proof of the general problem.

Figure 2-7 shows the inductance and Q_u of a close-wound 20-turn coil of No. 20 Formvar wire based on center and end placement on the same test rod. Moving the coil from center to the rod end caused an inductance decrease of 46% and a Q_u reduction of 47%.

There is still another consideration with respect to the coil geometry and the ferrite rod. The inductance for a coil wound tightly on the core will be considerably higher than when the coil is spaced away from the core. These effects are highlighted in Figs. 2-8 and 2-9. An investigation was conducted with a 4-in. ferrite rod, 0.5 in. in diameter, the μ_i of which was 125. By

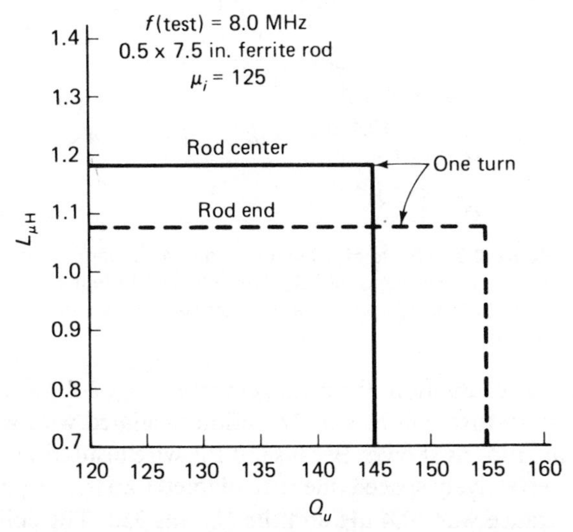

Figure 2-6 Graphic illustration of Q versus inductance for one turn of wire at the end and center of a ferrite rod.

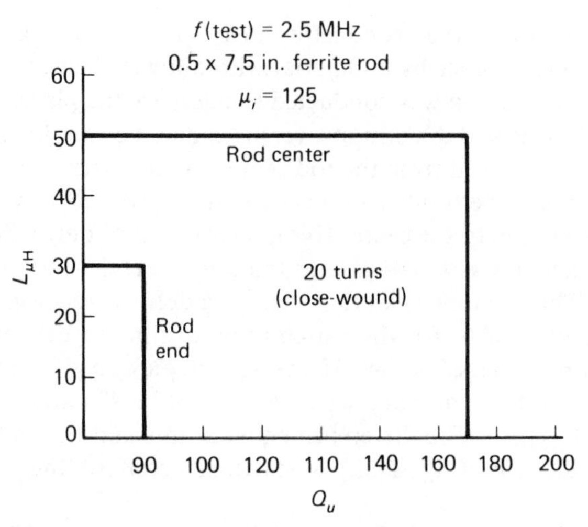

Figure 2-7 Comparison between the Q and inductance of a 20-turn coil when moved from the center of a rod to the end of the rod.

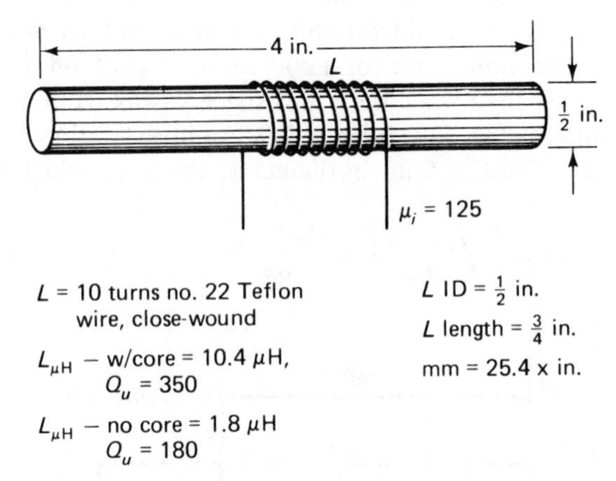

Figure 2-8 Effects of a ferrite rod on an air-wound coil that has an inductance of 1.8 μH. Insertion of the ferrite rod raises the inductance and the Q. The coil is tightly wound on the rod.

means of Q-meter evaluation, the arrangement of Fig. 2-8 was investigated. A close-wound, 10-turn coil of No. 22 Teflon-insulated wire was placed at the center of the rod, as shown. Because of the wire insulation, the conductor turns were effectively spaced one wire diameter apart, or nearly so. The resultant inductance was 10.4 μH and the Q_u was 350. The coil inductance without the magnetic core present was 1.8 μH.

Next, a coil was wound with the same type of wire, close-wound, to a

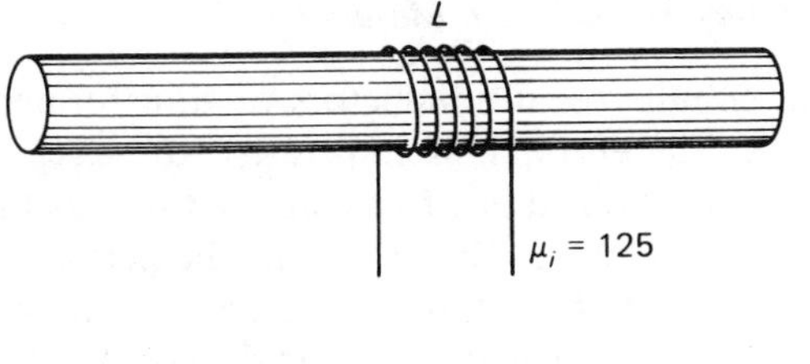

L = 6 turns no. 22 Teflon wire, close-wound

$L_{\mu H}$ – w/core = 4.6 μH
Q_u = 190

$L_{\mu H}$ – no core = 1.8 μH,
Q_u = 180

L ID = $\frac{3}{4}$ in.

L length = $\frac{3}{8}$ in.

Coil-to-rod spacing = $\frac{1}{8}$ in.

Figure 2-9 Conditions similar to those of Fig. 2-8 except that the air-wound coil is not wound tightly on the rod. Rather, an air gap of 1/8 in. exists between the coil and the core. The inductance and Q increases, but not as markedly as when the coil is tight against the core material.

¾-in. diameter. It was configured to an inductance of 1.8 μH without the core inside it. With this inductor located centrally on the rod, spaced ⅛ in. from the rod (coaxially), the measured inductance was 4.6 μH—somewhat less than in the example of Fig. 2-8. Furthermore, the Q_u degraded to 190. Figure 2-9 shows the placement of the coil, L. In both examples, the initial 1.8-μH coils exhibited an unloaded Q of 180. The coil of Fig. 2-9 required only six turns of wire for an inductance of 1.8 μH, owing to the larger diameter. Since the ¾-in. diameter coil was necessarily shorter in length than the ½-in.-diameter one of Fig. 2-8, the l_r/l_c effects were present and must be taken into account, as shown by the Polydoroff equation. However, for the purpose of comparison the results are ample for demonstrating the reduction of Q and inductance as the coil is spaced away from the rod.

As a result of numerous laboratory tests the author concluded that the best values of Q_u are obtained when the coil turns are spaced one wire diameter apart rather than being close-wound, when the coil is lumped at the center of the rod. The graph of Fig. 2-4 shows clearly that spacing the coil turns over the entire length of the rod achieves excellent Q_u characteristics also, offering slightly higher values of Q than when a shorter coil with spaced turns is lumped centrally on the rod.

Tests conducted with Litz wire versus solid, enameled magnet wire indicated that the Litz wire provided superior Q_u values in all instances. However, there seemed to be little difference between the results when No. 22 stranded wire with Teflon insulation was compared to Litz wire of comparable gauge.

2.1.3 Frequency versus Core Material

Narrow-band circuits require inductors or transformers which exhibit high orders of Q. Therefore, it is imperative to select the proper core material for the application. It is a fortunate fact that as the operating frequency is increased in a narrow-band circuit, the permeability of the core material must be lowered. Were this not true, a small inductance, such as 1 μH, might require less than one wire turn on the core used, assuming that a relatively high μ_e predominated. The consequences would be degraded Q and impractical coil dimensions.

The notable exception to the foregoing discussion of core selection versus operating frequency is in the case of broadband transformers. In that type of application it is quite desirable to use high-μ cores, because as the operating frequency is increased, the core permeability effectively "disappears," whereas at the lower end of the chosen frequency spread the core μ_e ensures ample inductance. This subject will be treated in greater depth when we examine broadband transformer concepts. In this chapter we will address the subject of narrow-band circuits for the most part.

Some of the significant properties of various ferrite core materials are listed in tabular form in Table 2-1. Of particular significance to this discussion is the column that lists the recommended frequency ranges of the materials. The permeabilities contained in the table are standard ones offered by Fair-Rite Products Corp. and are similar to those of other manufacturers. It can be seen that as the recommended operating range is elevated, the μ_i of the core material becomes lower. This illustrates why inattention to core characteristics can lead to unworkable Q values. These same general rules apply to powdered-iron cores also.

TABLE 2-1 Ferrite-core characteristics

Initial Permeability, μ_i	*Maximum Permeability, μ_m*	*Saturation Flux Density, B_s, at 13 Oe*	*Recommended Frequency Range[a] (MHz)*	*Fair-Rite Material*
20	—	2000 at 40 Oe	80-100	68
40	—	3000 at 20 Oe	10-80	67
100	370	2150	5-15	65
125	450	2350	0.2-10	61
175	400	2550	0.1-5	62
250	375	2200	0.05-4	64
300	3600	3900	0.001-5	83
800	1380	2500	0.01-1	33
850	3000	2750	0.01-1	43
1200	2500	3400	0.01-1	34
1800	4600	1150	0.001-2	77
2000	3500	3500	0.001-1	72

Courtesy of Fair-Rite Products Corp.
[a]Frequency ratings for optimum Q in narrow-band tuned circuits.

2.2 Rods and Bars as Antennas

Ferrite rods and bars are used in numerous loop-antenna applications such as broadcast-band receivers, low- and medium-frequency direction-finder receivers, in the receivers used for airborne navigation, and even in pocket-size VHF receivers. A closed-core magnetic device cannot serve the same purpose as the rod or bar when used as a loop-antenna core. This is because the toroidal configuration is self-shielding, thereby preventing the desired signal pickup. Similarly, there is a degrading effect in rod-antenna efficiency when the length versus diameter *(l/d)* is carried beyond approximately 35. As the length is increased past that point, the "toroidal effect" becomes more and more pronounced. In this context an infinitely long ferrite or powdered-iron rod becomes equivalent to a closed circuit at full permeability.

The electrical parameters of a magnetic-core loop antenna are roughly equivalent to those of the early-day air-core loop or "frame" antenna seen in Fig. 2-10, which was developed by Hertz in 1888. It was known then as the "loop collector." Various forms of this type of antenna have been used since for reception of broadcast-band signals, navigational signals aboard vessels and aircraft, and amateur-radio signals. The air-core loop can be used as an open system or as an electrostatically shielded device, as seen in Fig. 2-11. In the latter form, the loop discriminates various types of noise and corona static. Ferrite-rod loops can also be equipped for electrostatic shielding. This subject will be covered later in the chapter.

Loop antennas, whether the air-core or magnetic-core variety, have a height factor, specified as h. This is called the effective height, h_e. This is expressed in meters as a factor, which, multiplied by the field strength in μV per meter provides the loop-induced voltage in μV:

$$h_e = \frac{2\pi NA}{\lambda}$$

where $d/\lambda<1$, N is the number or loop turns, $\pi = 3.14$ and A is the area of each turn in square meters. The effective height obtained from the equation is for signals with horizontal polarity impinging on a loop that is orientated for maximum response.

The magnetic-core loop exhibits an interesting and superior trait when compared to an air-core loop: The h_e increases as a function of μ_e. Thus,

$$h_e = \frac{2\pi NA\mu_e}{\lambda}$$

where, for practical considerations, λ is in centimeters and A is in cm^2.

During laboratory and outdoor test-site investigations, the author observed that signal pickup over an efficient air-core loop is somewhat less

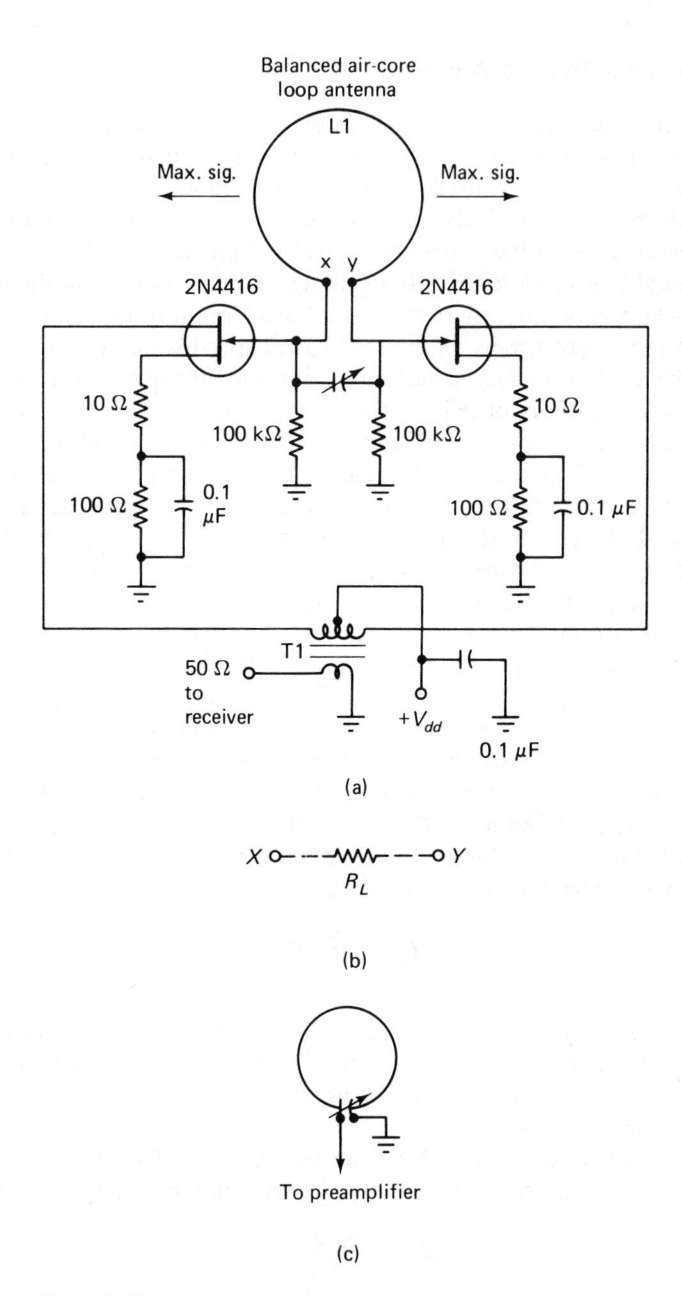

Figure 2-10 Example of an air-core loop or frame antenna (a). A balanced preamplifier will ensure minimum pattern distortion. R_L, as seen by the preamplifier, is depicted in (b). It is normally on the order of 1 Ω or less. The feed system in (c) is sometimes used, but imbalance will cause pattern skewing and poor nulling.

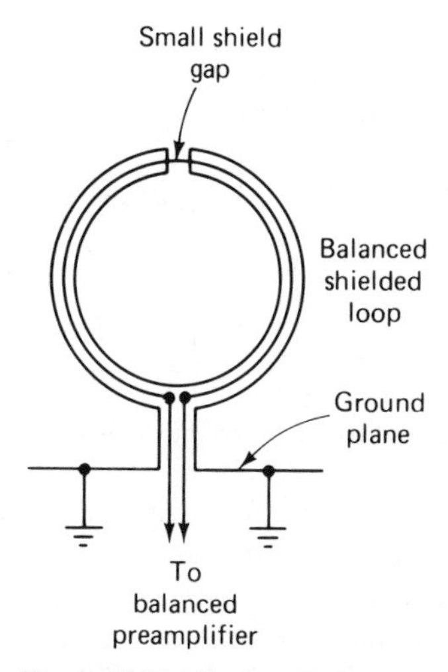

Figure 2-11 Format for a balanced, magnetically shielded loop antenna.

than μ_e with very short rods (less than $l/d = 5$). But, when the ratio is greater than 5, the full benefits of μ_e are realized.

Tests were conducted with lumped turns at the center of rods, at the ends of the cylindrical cores, and with the coil turns spread uniformly over the entire rod. Irrespective of the resultant Q values obtained from the three formats, the greatest h_e was noted when the turns occupied the entire core length. This is because the maximum number of turns was necessary to secure the required inductance, as seen in Fig. 2-4. These results are compatible with the equations for h_e, where a significant term is N. These evaluations were performed with cylindrical ferrite cores which had an l/d ratio of 30 (0.5 × 15-in. rod), which was obtained by joining two 7.5-in. rods, end to end, by means of epoxy cement. The μ_i of the material was 125.

The test method of Fig. 2-12 was employed during the tests. The signal source was a commercial AM broadcast station, located 1 mile from the test site. No conductive objects were closer than ¼ mile from the loop under test with respect to the signal path. The loop was oriented for maximum figure-eight response for each test.

An air-core frame loop of comparable area to the 0.5 × 15-in. rod loop was fabricated and tested as in Fig. 2-12. The magnetic-core loop with spread turns exhibited a gain of approximately 50 over the air-core version. However, when the rod loop was compared to a free-standing vertical antenna with buried radials, the loop was some 30 dB less effective when oriented for the same polarity. The vertical was ⅛ wavelength high but ad-

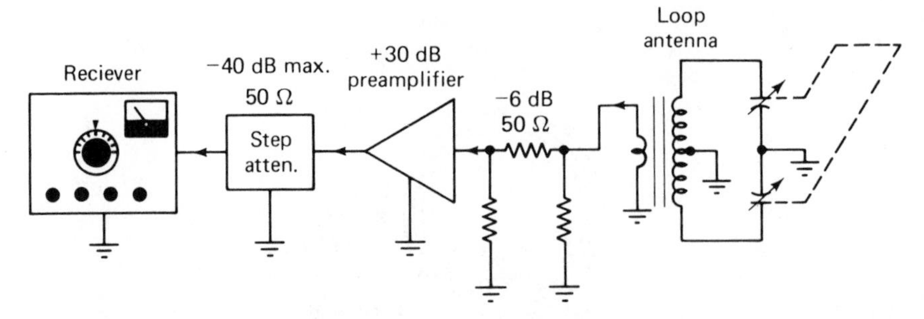

Figure 2-12 Test method used by the author during evaluation of a ferrite rod antenna.

justed electrically to ¼ wavelength by means of top loading. The feed point was matched for 50 Ω at the operating frequency.

A transmitting loop can be used as a signal source for more precise testing. In an exercise of this type, current in the transmitting loop is imposed by a single generator. For spacings of $\lambda/2$ or less, the field strength is μV per meter at the test loop is

$$E = \frac{18.85Nr^2I}{X^3}$$

where r is the transmitting loop radius, N the number of turns, I the current in mA, and X the spacing between the two loop antennas in meters, as discussed by Polydoroff[1]. He adds that the results can be expressed alternatively by

$$E = \frac{1{,}180Nr^2e}{X^3fL}$$

where with regard to the transmitting loop, e is the voltage across the loop in μV, f the frequency, and L the inductance of the transmitting loop in μH. In this case E is the voltage across the receiving loop, as in the previous equation.

2.2.1 Loop Pattern Symmetry

The primary purpose of loop-antenna utilization, apart from effecting a physically small component, is to take advantage of the nulls in response which exist off the sides of the classic figure-eight response pattern. The characteristic minima and maxima are useful in radio direction-finding applications. This need dictates a requirement that the pattern be as symmetrical as possible (Fig. 2-13a). In this context an ideal response produces a pair of zero points that are precisely 180° apart. If the disturbances in symmetry are only minor, a pattern distortion of many degrees can result (Fig. 2-13b).

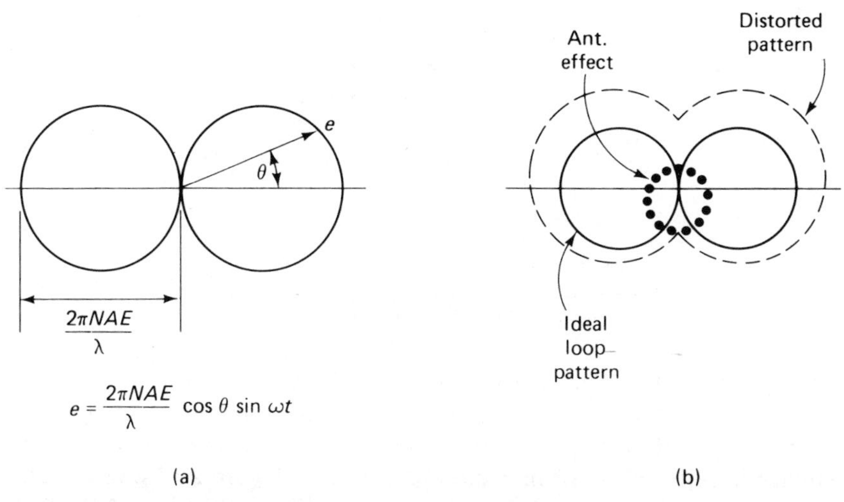

Figure 2-13 Classic response pattern of a properly balanced loop antenna (a). In (b) the pattern can be distorted many degrees by only a slight imbalance in the system.

An ideal loop should serve entirely as the antenna, and in accordance with its intended characteristics. Imbalance in the physical layout (excessive asymmetry, long connecting leads, etc.) can lead to what is known as *antenna effect.* In such an example the conductors apart from the main body of the loop will act as unwanted parts of the antenna and will respond to incoming energy differently than the loop will. This is represented by the circle with dots in Fig. 2-13b. Other factors can degrade the ideal response pattern, such as displacement of current. This can result from excessive axial length in a loop antenna. It is important, therefore, that close attention be paid to the physical attributes of a loop antenna. Also, balanced loops will yield superior results to those which do not feature electrical balance (Figs. 2-10 and 2-11). Electrostatic shielding of the type illustrated in Fig. 2-11 is extremely helpful in reducing the antenna effect. The major limitation in using shielded loops is that the loop must be physically close to its ground plane for best results. Electrostatic shielding of a rod or bar type of loop can be carried out by locating the loop at the physical center of a U-shaped channel of aluminum. This is shown in Fig. 2-14. Most portable direction-finder receivers utilize this method. Test-range results indicate no degradation in loop efficiency when the shield is added, provided that the top opening of the U channel is kept entirely open.

2.2.2 Noise Immunity

Loop antennas are considerably better than vertical antennas with respect to noise immunity. Weak signals that are buried in the noise to which a vertical antenna is highly responsive are not capable of conveying

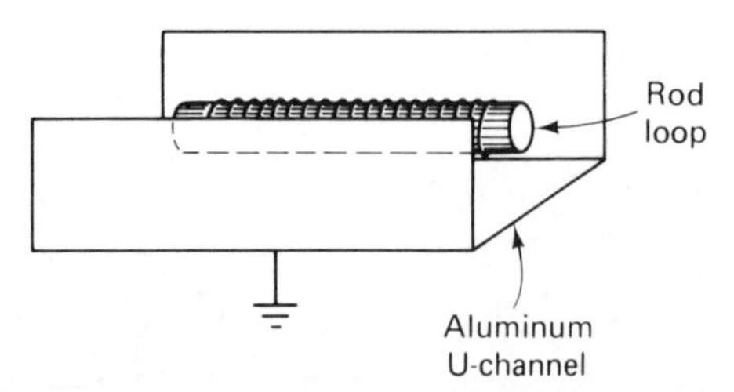

Figure 2-14 Method for providing magnetic shielding of a ferrite-rod loop antenna. An aluminum U-channel is employed.

intelligence reliably. The high-Q properties of a well-designed loop antenna (narrow bandwidth) are not as responsive to noise as is true of low-Q loops or other types of antennas. Furthermore, the loop nulls can be used to advantage in reject noise from a specific direction. Figure 2-15 gives a comparison in signal-to-noise ratio between a low-Q loop and one that has high Q. The advantage of designing for high Q is readily apparent from examination of the two response curves. The trade-off is, of course, in bandwidth. A fixed-tuned loop would not be especially useful if a frequency excursion beyond f^1 and f^2 were required. In situations where the loop is continuously tunable over the desired frequency range, or where it can be pretuned to a separate frequency and left in that state, undercoupling to the load is recommended in the interest of improved noise ratio. A low-noise preamplifier is a simple and economical way to compensate for the loss brought about by light load coupling. The higher Q_L (loaded Q) can be very advantageous.

2.3 Practical Loop Circuits

Some assembled magnetic-core loops are shown in Fig. 2-16. Most of the commercially built loops employ Litz wire for the purpose of elevating the circuit Q as much as possible. However, the mere use of that preferred type of wire does not ensure optimum Q. The results of using magnet wire can be seen by examining Fig. 2-17, where Q_u (unloaded Q) measurements were made on a short bar type of loop which was manufactured in the Orient. A sample A_L calculation indicates the μ_i of the material to be on the order of 250, which may account in part for the midrange Q value. The winding consists of small-diameter (No. 28 enameled) wire. Although the Q_u is not spectacular, it is adequate for most pocket-size AM radios—the type from which the loop was taken. Most small AM radio loops are overcoupled to the load (the mixer stage) to minimize loss. Were an RF amplifier employed, lighter coupling would be practical, and the Q_L would be much higher. The benefit of higher Q would be seen in improved receiver front-end selectivity

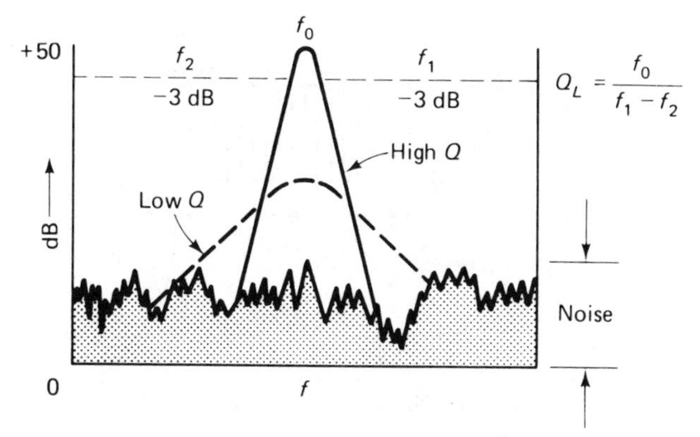

Figure 2-15 Comparison of signal-to-noise ratio for a low-Q and a high-Q loop antenna. The trade-off is in bandwidth.

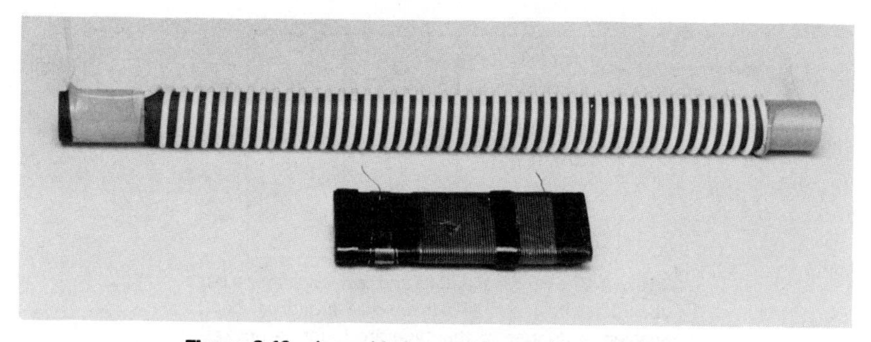

Figure 2-16 Assembled magnetic-core loop antennas.

. . . an aid in separating strong broadcasting stations in metropolitan locations.

The turns of section *B* in Fig. 2-17 are spaced across roughly two-thirds of the bar, with the turns of section *A* being close-wound. This is a common scheme in foreign-made radios, presumably to allow section *A* to be used for final inductance trimming by spreading the turns slightly. The tap point between sections *A* and *B* is for connection to the input of a bipolar transistor mixer. The characteristic impedance of a mixer of that type is approximately 600 Ω at the base element. It can be seen from the tabular data in Fig. 2-17 that despite the *LC* ratio of the loop at any frequency in the broadcast band, the Q_e remains relatively constant. This enables the designers to use miniature variable capacitors in some radios, which have as little as 150 pF of maximum capacitance. The increase in the required inductance under those conditions makes it necessary to use more turns of wire in the loop. The advantage of this was discussed in Sec. 2.2. It is likely that the loop of Fig. 2-17 would have exhibited somewhat higher Q if Litz wire had been used

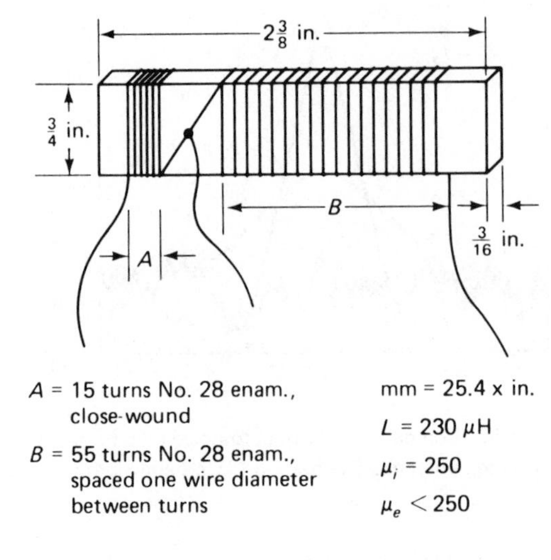

Frequency (kHz)	Q_u	C_{PF}
550 (kHz)	180	360
800 (kHz)	175	175
1000 (kHz)	180	110
1600 (kHz)	180	40

Figure 2-17 Physical details and test results for a bar type of ferrite loop used in an AM broadcast-band receiver. Antennas of this type are found in pocket-size radios built in the Orient.

instead of No. 28 magnet wire. But, when pocket-size radios are built for retail at less than $10, a trade-off between performance and economy is the rule more often than not.

2.3.1 Loop-to-Mixer Circuits

An example of how a ferrite-rod loop can be used directly into the mixer stage of a broadcast-band receiver is given schematically in Fig. 2-18. The rod characteristics and winding information for this example and the one in Fig. 2-19 are supplied in Fig. 2-20. The circuit of Fig. 2-18 shows the loop with a tap near one end. The tap is set for approximately 600 Ω to match the characteristic input impedance of the bipolar transistor mixer, Q1. With the specified loop inductance, 240 μH, the entire broadcast band can be covered by means of a single-section 365-pF variable capacitor. In a typical design such as one finds in low-cost AM radios, the benefits of the high Q_u are negated to a greater extent by the tight coupling to the mixer.

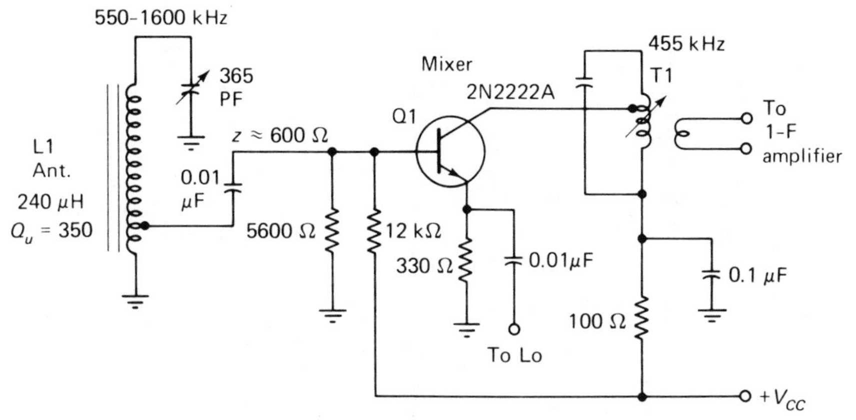

Figure 2-18 Typical example of a ferrite-rod antenna as used with the mixer of an AM broadcast receiver. The mixer base is tapped at approximately 600 Ω on the rod antenna.

An alternative coupling method between the loop and Q1 would be to employ a secondary winding, or link, in place of the tap on L1. If a Q_L versus gain trade-off can be used without too much sacrifice in receiver sensitivity, the coupling link can be made small or the tap on L1 can be located closer to the grounded end of the loop.

A method for using a loop antenna with a dual-gate MOSFET mixer is provided at Fig. 2-19. A 3N211, 40673, or any of the similar high-frequency MOSFETs would be suitable in this circuit. The signal gate is tapped down partway on the loop, L1, to enhance circuit stability. This technique lowers the input seen by the gate impedance of the FET progressively as the tap is moved toward the ground end of L1. The most likely manifestation of mixer self-oscillation is experienced when the receiver is tuned to the low-frequency end of the broadcast band. This places the FET Input at a frequency near the 455-kHz IF, thereby encouraging tuned gate/tuned drain oscillation. Strapping gate 1 nearer to the cold end of L1 eliminates the problem. Since gate 1 of the FET is greater than 1 MΩ in impedance, the mixer does not degrade the Q_u of the loop significantly. This means that a loop used with a FET mixer is capable of providing much greater receiver front-end selectivity than would be likely with the circuit in Fig. 2-18. Furthermore, the FET mixer is capable of much better front-end dynamic range—an objective of many receiver designers.

2.3.2 Loop to RF Amplifier

It is a simple matter to resolve the front-end selectivity problem which is so common in low-cost broadcast-band receivers. For a slight cost increase, the designer can incorporate a low-noise RF amplifier ahead of the mixer, as illustrated in Fig. 2-20. This stage can be used to compensate for the loss

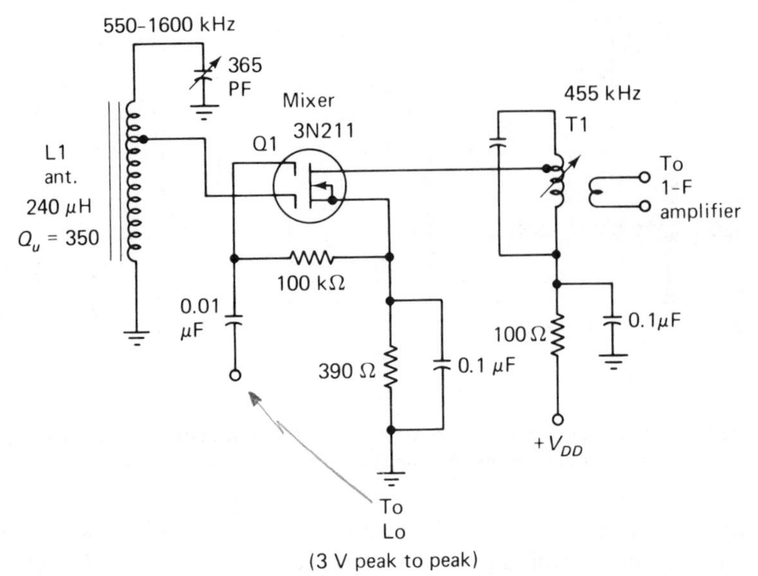

Figure 2-19 Technique for using a MOSFET mixer with a rod antenna. The FET gate impedance is on the order of 1M Ω or greater. This permits placement of the tap near the high-impedance end of the rod antenna.

that results from undercoupling the loop to its load. This allows L2 to be set for the desired Q_L or loop BW_L (loaded bandwidth).

Transistor Q1 in this circuit is configured as a fed-back, broadband amplifier. The characteristic input and output impedance of the stage is approximately 50 Ω. A combination of degenerative feedback (R1) and shunt negative feedback (R2) ensure unconditional stability of Q1, even when very light coupling is used between the loop and the amplifier. A 4:1 broadband toroidal transformer, T1, steps the 200-Ω down to 50 Ω. T1 contains 20 bifilar turns of No. 28 enameled wire on a 0.5-in.-diameter ferrite toroid core. The μ_i of the core is 950. A 0.37-in.-diameter core of the same μ would be suitable if smaller-gauge wire were employed. The 2N5179 transistor was chosen for its high f_T rating (> 1000 MHz) and low-noise characteristic. Typical gain is 15 dB. Another advantage of this type of circuit is that the 50-Ω output of the RF amplifier is compatible with diode-ring doubly balanced mixers, the ports of which are also 50-Ω. The gain of Q1 would ensure adequate receiver noise figure if a DBM (double balanced mixer) were used. Such a mixer would greatly elevate the receiver dynamic range. The port-to-port isolation of the diode-ring mixer, plus the inclusion of RF stage Q1, can virtually eliminate unwanted local-oscillator radiation via the antenna.

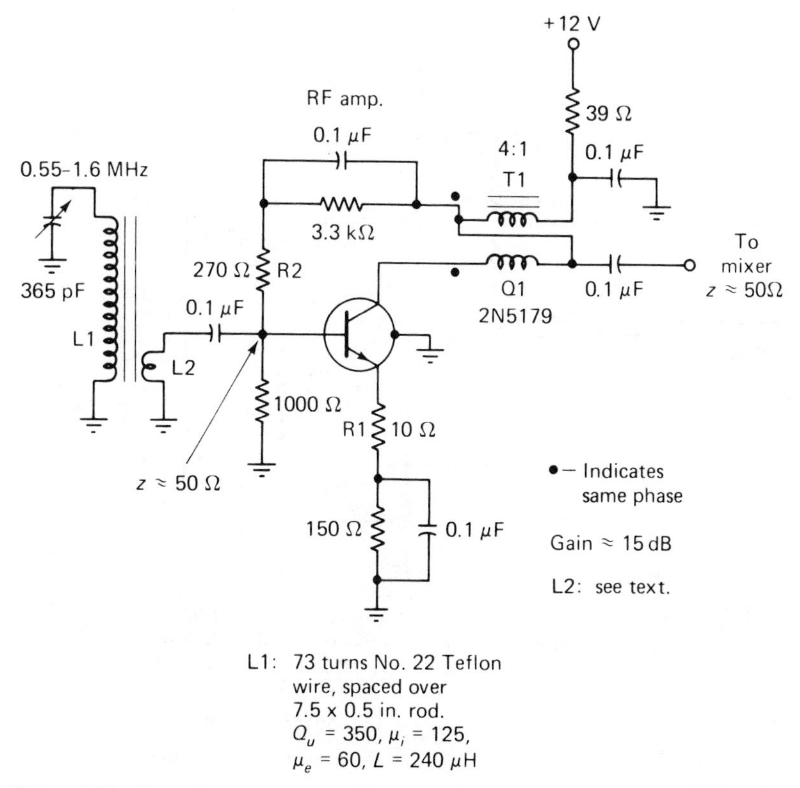

Figure 2-20 Example of a fed-back bipolar-transistor RF amplifier as used with a rod loop antenna. Shunt and degenerative feedback are used to ensure unconditional stability.

2.3.3 Loop with Sense Antenna

A practical circuit for a ferrite-rod loop antenna used in combination with a sense antenna is shown in Fig. 2-21. An arrangement of this variety is useful when a cardiod response pattern is desired. A relatively unidirectional pattern, which is heart-shaped, as seen in Fig. 2-21b, is needed in direction-finding equipment to avoid ambiguity of compass headings. A bidirectional pattern can confuse an operator, especially if the operator is past the radiobeacon transmitter being sought for navigational purposes.

The sense antenna consists of a short vertical element placed close to the magnetic-core loop—typically from 6 to 36 in. in front of the rod or bar loop. The sense-antenna voltage is combined and placed in phase with the loop voltage by means of R1 of Fig. 2-21a. The in-phase relationship results in the cardiod pattern in Fig. 2-21b. The factor k is equal to the fractional

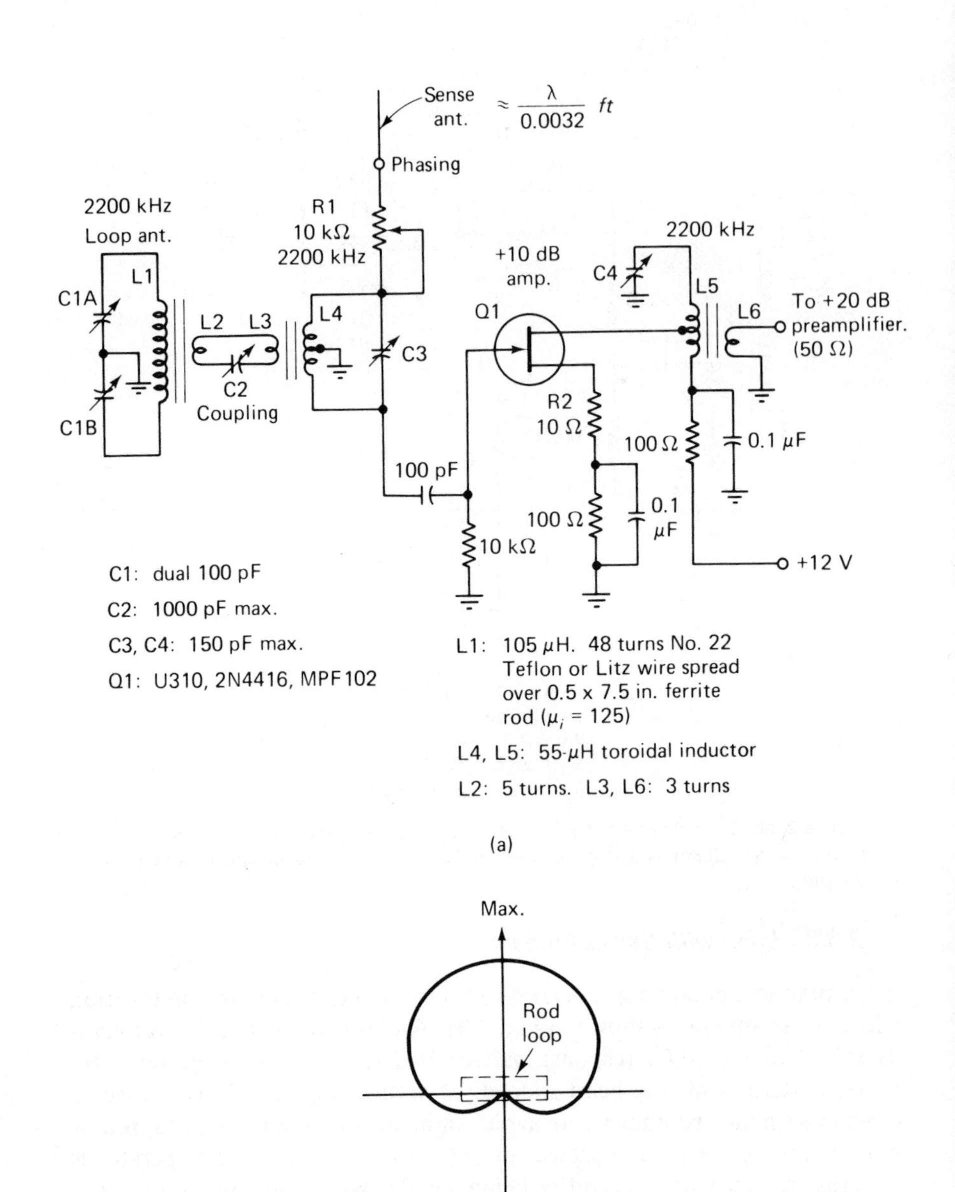

Figure 2-21 Circuit of a ferrite-rod loop antenna with sense antenna and preamplifier (a). The resultant cardioid response pattern is shown in (b).

relationship of the sense antenna to loop maximum voltage. A cardioid response results when $k = 1$. If k were changed to 0.5, a small back lobe would protrude. By reducing k to 0.2, a figure-eight pattern would develop as the loop degraded toward the classic bidirectional characteristic found when no sense antenna is used.

A backside null depth as great as 40 dB has been measured by the author on a test range while using the circuit of Fig. 2-21a. This was realized only after repeated adjustment of R1 and C1 for the best null depth: There is considerable interaction between the two adjustments. Null depths that were measured off the ends of the rod loop without the sense-antenna circuitry were on the order of 23 dB. In theory, and with perfect loop balance, the two nulls could exceed 30 dB. This was explained to the writer by Henry Jasik during a discussion on loop antennas in 1975 [2].

Loop balance in the circuit given in Fig. 2-21a is provided by using a split-stator tuning capacitor, C1A and C1B. C2 is used to adjust the coupling between the loop and the load, consistent with the desired Q_L characteristic. The circuit from L2 to L6 is isolated from the loop circuit by means of shielding to prevent stray coupling between the circuits. The length of the sense antenna may have to be adjusted empirically to obtain the antenna voltage required. For the frequency specified, an 18-in. length of conductor proved entirely adequate.

A low-noise post-loop amplifier, Q1, is used to ensure an acceptable noise figure. Degenerative feedback is made available by the unbypassed resistor, R2. This, plus tapping the FET drain down on L5, prevents self-oscillation of the RF preamplifier. These measures lower the stage gain from a possible 20 dB to roughly 10 dB. The loop can be shielded electrostatically if desired. The details were given in Fig. 2-14.

2.3.4 Loop Variations

When it becomes necessary to make a loop antenna variable in terms of inductance, two common approaches can be taken. These are seen in Fig. 2-22. The need for varying the inductance is encountered when the RF amplifier or mixer stage of a receiver must track with the local oscillator. The employment of a two-section variable capacitor (ganged) makes this almost mandatory.

Figure 2-22a shows a compromise type of loop antenna, used primarily as a replacement unit for AM radio loops that have become damaged. This style of adjustable loop is short—usually 2 to 3 in. in length. A magnetic-core slug is adjustable from one end to obtain the required inductance. The Q of this antenna is relatively low compared to long rod or bar loops . . . in the area of 150. To enhance the pickup capability of the loop, a short length of wire (1 to 3 ft long) is sometimes connected to the ungrounded end of the coil and allowed to dangle outside the receiver cabinet. The

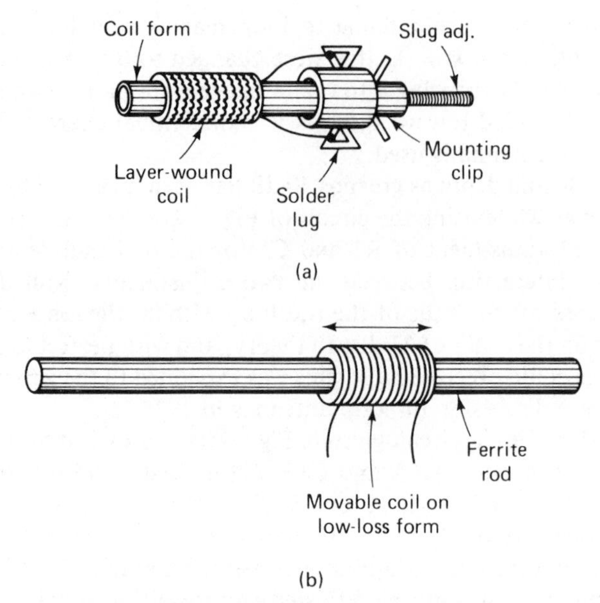

Figure 2-22 Small ferrite loop antenna (adjustable inductance) is seen in (a). A long movable ferrite rod is contained inside a layer-wound coil. The coil inductance in (b) can be adjusted by sliding it along the ferrite rod.

directional characteristics of this type of loop are practically useless due to physical and electrical asymmetry.

When space permits, the technique shown in Fig. 2-22b is superior to that in (a). The loop coil is wound on a movable insulating form. Tracking is achieved by moving the coil left or right until the desired inductance results. The coil is then cemented in place. The lumped inductance is not as effective as that which is spread over the entire length of the ferrite rod, but acceptable results can be had with this general scheme.

2.4 Rod-Core Transformers

Although ferrite rods are used primarily as cores for loop antennas, they are employed sometimes in broadband transformers. More typically, however, toroids and pot cores are chosen as the core material for narrow- and broadband transformers.

The major disadvantage in using a solenoidal winding on a rod core is that the inductor or transformer lacks the self-shielding characteristic of the toroid or pot-core equivalent. Apart from that shortcoming, the rod is an entirely suitable magnetic-core material. It can be used as the foundation element in narrow- or broadband circuits. Some manufacturers prefer the rod format over that of the toroid because of reduced winding complications, especially if a toroid-winder machine is not available.

Figure 2-23 portrays in pictorial and schematic form a broadband 4:1 transformer as it would be configured on a ferrite rod. The rules given in the chapter on transformer design apply in this example, even though a closed core is not utilized. Although the wire pair is shown in a parallel placement format, some designers prefer to twist the wires together before laying them on the core. Normally, 6 to 10 twists per inch is the common rule.

2.5 Slug Applications

Slugs are similar in structure to magnetic-core rods. The essential difference is that they are smaller in size than loop-antenna rods. Also, they are manufactured with some type of physical adjustment scheme in mind.

Slugs are used as movable elements inside coil forms that are made of

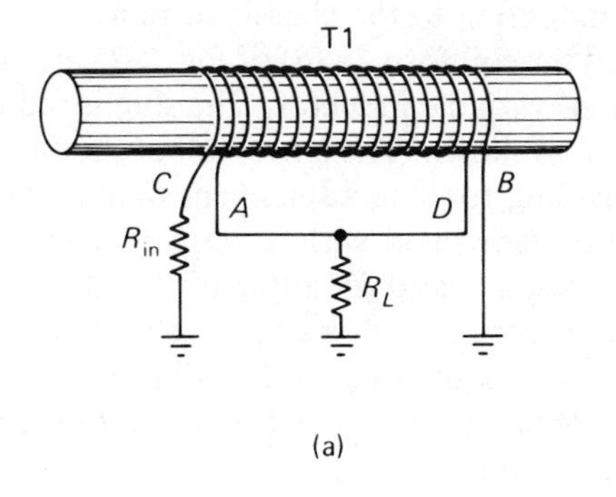

(a)

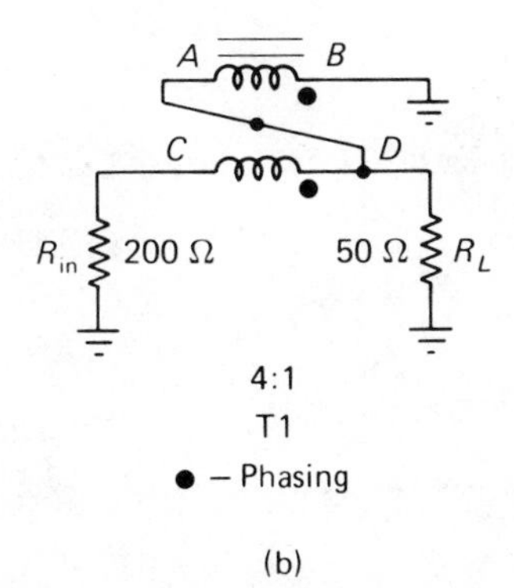

(b)

Figure 2-23 Pictorial (a) and schematic (b) illustrations of a ferrite rod as used in a 4:1 broadband transformer.

ceramic, steatite, phenolic, plastic, or paper. The composite assembly is called a slug-tuned inductor or transformer, once the winding is in place. The slug material can be ferrite, powdered iron, or even brass. The latter is often used at VHF and UHF to effect changes in inductance. Brass slugs have the opposite effect of ferrite or iron cores. Brass *decreases* the coil inductance as it is moved deeper into the coil. The opposite is true of the other two materials.

Figure 2-24 shows an assortment of manufactured slug-tuned inductors in photographic form. It can be seen that they are available in a host of sizes and shapes. Some are intended for mounting on printed-circuit boards, while others can be affixed to metal chassis or panels.

Two of the most common formats for slug-tuned coils are illustrated in Fig. 2-25. The slug in (a) is for use with the coil form in (c). A screw is embedded into the core material during the manufacturing process. The end of the screw is slotted, as is the bottom end of the slug. This permits adjustment from either end of the coil form. This style of slug-tuned coil is used primarily for direct mounting to the chassis or panel.

The slug in (b) and the coil form in (d) of Fig. 2-25 are combined to form a printed-circuit-mount variable inductor. The slug is molded with threads on the outside surface, as shown. The inner wall of the coil form has mating threads, thereby providing for slug adjustment in and out of the coil. The slug is either slotted or fabricated with a hex hole completely through its center. The hex or the slot are used for adjusting the slug positioning. If further adjustment is not contemplated, wax can be melted into the top or bottom of the coil form to affix the slug so that vibration does not change the setting. A jam nut or locking spring can be used with the coil form in Fig. 2-25a and c for the same purpose.

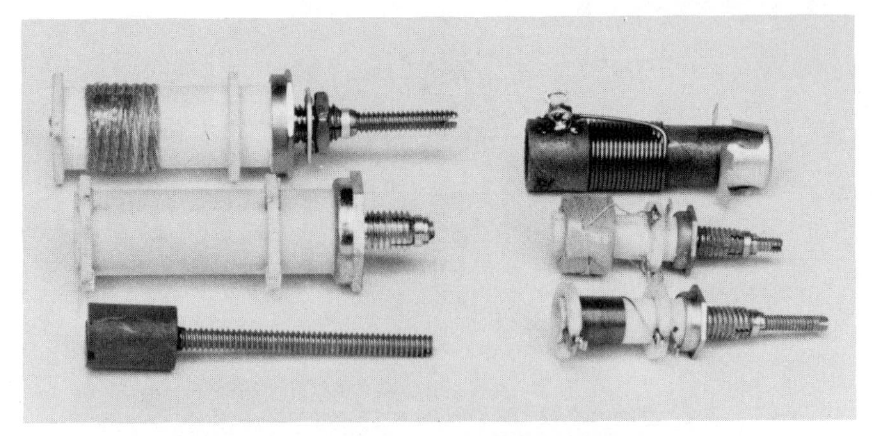

Figure 2-24 Assortment of typical slug-tuned inductors.

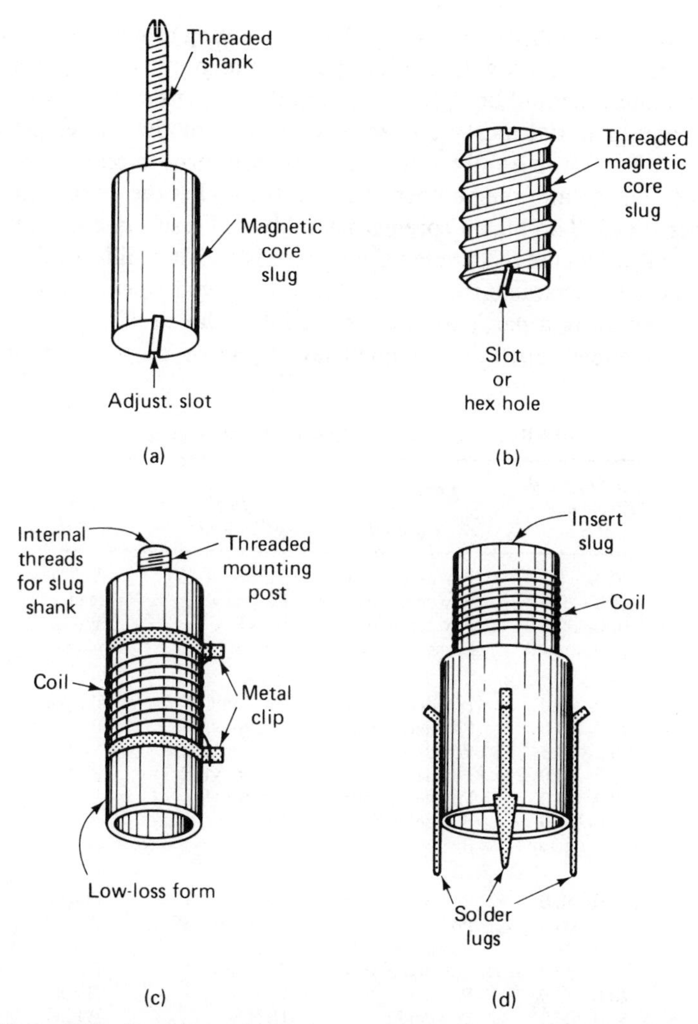

Figure 2-25 Breakdown views of two popular slug-tuned inductors. The slug in (a) mates with the coil form in (c). Similarly, the treaded slug in (b) mates with the printed-circuit style of coil form in (d).

2.5.1 Performance Considerations

Slug-tuned inductors or transformers are almost without exception employed in narrow-band circuits. This dictates a need for specific criteria, notably electrical stability, mechanical stability, and circuit Q. A given resonator, once adjusted, should remain in the intended state within reasonable constrictions. This is irrespective of temperature changes and vibrations.

Mechanical stability requires that the coil winding be held firmly in place on the coil form with low-loss coil dope, such as polystyrene cement. The slug mechanism should be tight in its threaded collet or coil form.

The core material must be selected correctly for the chosen operating frequency. Too low a μ will require excessive coil turns to realize a given inductance. The greater the number of turns, the higher the dc resistance and the lower the Q. Too high a core permeability will result in a small number of turns but will cause a degraded Q characteristic. Thus, it is prudent to examine the core characteristics versus recommended operating frequency before committing a particular core material to the circuit. Table 2-2 lists the recommended operating frequencies for most of the popular core

TABLE 2-2 Screw-shaft slug characteristics.

Recommended Frequency MHz	*Core Material*	*Basic Powder*	*Color Code*
0.2- 1.5	Powdered iron	Carbonyl C	Yellow
1.0- 20.0	Powdered iron	Carbonyl E	Red
50.0-200.0	Powdered iron	IRN 8	White
100.0 and up	Brass		None
2.0- 40.0	Powdered iron	Carbonyl TH	Purple
40.0-300.0	Powdered iron	Carbonyl SF	Blue
0.2- 1.5	Powdered iron	Carbonyl C	Yellow
1.0- 20.0	Powdered iron	Carbonyl E	Red
20.0- 50.0	Powdered iron	Carbonyl J	Green
50.0-200.0	Powdered iron	IRN 8	White
100.0 and up	Brass		None
2.0- 40.0	Powdered iron	Carbonyl TH	Purple
40.0-300.0	Powdered iron	Carbonyl SF	Blue
0.2- 1.5	Powdered iron	Carbonyl C	Yellow
1.0- 20.0	Powdered iron	Carbonyl E	Red
20.0- 50.0	Powdered iron	Carbonyl J	Green
50.0-200.0	Powdered iron	IRN 8	White
100.0 and up	Brass		None
2.0- 40.0	Powdered iron	Carbonyl TH	Purple
40.0-300.0	Powdered iron	Carbonyl SF	Blue
0.2- 1.5	Powdered iron	Carbonyl C	Yellow
1.0- 30.0	Powdered iron	Carbonyl E	Red
20.0- 50.0	Powdered iron	Carbonyl J	Green
50.0-200.0	Powdered iron	IRN 8	White
100.0 and up	Brass		None
2.0- 40.0	Powdered iron	Carbonyl TH	Purple
40.0-300.0	Powdered iron	Carbonyl SF	Blue

Courtesy of J.W. Miller Co., Division of Bell Industries.

materials available for slugs of the kind shown in Fig. 2-25a, as provided by J.W. Miller, a Division of Bell Industries. Table 2-3 gives the same general data for slugs of the type depicted in Fig. 2-25b. The core materials are brass or powdered iron. The recommended frequency ranges will ensure optimum Q. The Miller color code for slugs is included in the tables. The permeability factors are not listed in the tabular presentations, as the μ will depend on the core dimensions. The manufacturers of slugs can provide specific μ information versus the various slugs, with respect to size.

TABLE 2-3 Threaded-slug characteristics.

Recommended Frequency MHz	*Core Material*	*Basic Powder*	*Color Code*
0.2- 1.5	Powdered iron	Carbonyl C	Yellow
1.0- 20.0	Powdered iron	Carbonyl E	Red
20.0- 50.0	Powdered iron	Carbonyl J	Green
50.0-200.0	Powdered iron	IRN 8	White
2.0- 40.0	Powdered iron	Carbonyl TH	Purple
40.0-300.0	Powdered iron	Carbonyl SF	Blue
0.3- 1.0	Powdered iron	IRN 2	Yellow
.05- 5.0	Powdered iron	Carbonyl E	Red
4.5- 20.0	Powdered iron	Carbonyl TH	Green
20.0 and up	Powdered iron	Carbonyl J	White
100.0 and up	Brass		None
0.3- 1.0	Powdered iron	IRN 2	Yellow
0.5- 5.0	Powdered iron	Carbonyl E	Red
4.5- 20.0	Powdered iron	Carbonyl TH	Green
20.0 and up	Powdered iron	Carbonyl J	White
100.0	Brass		None
0.2- 1.5	Powdered iron	Carbonyl C	Yellow
1.0- 20.0	Powdered iron	Carbonyl E	Red
20.0- 50.0	Powdered iron	Carbonyl J	Green
50.0-200.0	Powdered iron	IRN 8	White
100.0 and up	Brass		None
2.0- 40.0	Powdered iron	Carbonyl TH	Purple
40.0-300.0	Powdered iron	Carbonyl SF	Blue

Courtesy of J.W. Miller Co., Division of Bell Industries.

It is important to consider the effect of the slug placement within a coil on the Q of the inductor. When the slug enters the coil completely, the Q is the highest within the recommended frequency range. Table 2-4 shows the results of laboratory tests made on two series 42 J.W. Miller coils. One is the μH range and the other is in the mH range. In both cases the Q drops as the slug is withdrawn from the coil. This effect clearly demonstrates the need to choose a coil which has the required Q_u (determined by the designer) at a specified value of inductance.

TABLE 2-4 Results of laboratory tests on J.W. Miller coils.

Slug Position	*L*	Q_u[a]	*Core Material*	*DC Resistance (Ω)*	*Test Frequency (MHz)*
Full	5.8 μH	155	Carbonal E	0.466	8
Half	3.8 μH	130			
Out	2.4 μH	100			
Full	5.8 mH	85	Carbonyl C	21	2.5
Half	3.8 mH	75			
Out	3.4 mH	70			

[a]Comparison between Q_u and position of the coil slug, indicating maximum Q when the slug is fully into the coil.

2.5.2 Practical Circuits

We have discussed the importance of Q when a slug-tuned inductor is used in a narrow-band circuit. Another significant point of concern is stability versus temperature. The effects of long-term drift are particularly annoying in most free-running oscillators of the type shown in Fig. 2-26a. Typically, drifts in frequency which exceed, say, 100 Hz are unacceptable by present-day performance standards. In stringent applications the drift cannot be permitted to go beyond a few hertz.

In an ideal situation the designer would not use an inductor for frequency control if that inductor contained a magnetic core. This is because the core permeability changes with temperature. Alternatively, the designer would attempt to utilize a mechanically rigid coil which was air wound or contained on a low-loss form such as ceramic. The shortcoming of this approach is brought on by today's emphasis on miniaturization. A good air-wound, stable inductor for the high-frequency spectrum and the lower would occupy considerably more volume than a magnetic-core component which provided equivalent inductance.

We may ask: What alternatives are there with regard to stability? The classic procedure when magnetic cores are contained in the oscillator tank coil is to use temperature-compensating devices to correct for long-term drift. Most core materials have a positive drift characteristic—increasing μ_e with an increase in temperature. As this happens, the Q degrades somewhat. Carbonyl E powdered-iron mix, for example, will have a μ_e change of roughly +0.005%/°C. The Q will degrade by approximately −0.02%/°C. The change in permeability with heat can be corrected by using fixed-value capacitors with negative-coefficient properties.

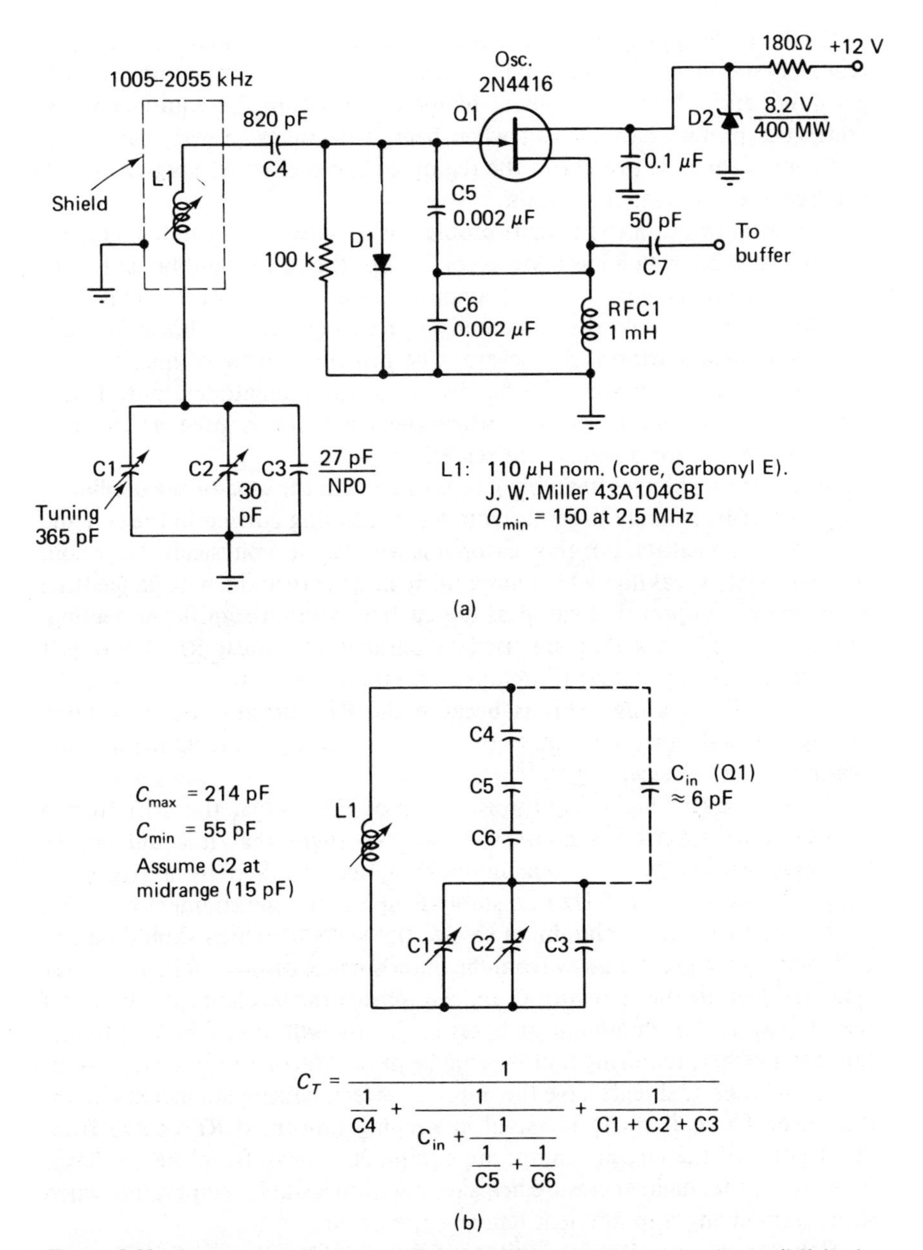

$$C_T = \cfrac{1}{\cfrac{1}{C4} + \cfrac{1}{C_{in} + \cfrac{1}{\cfrac{1}{C5} + \cfrac{1}{C6}}} + \cfrac{1}{C1 + C2 + C3}}$$

(b)

Figure 2-26 Design example for a VFO that contains a slug-tuned inductor (L1) in the frequency-determining part of the circuit. Excellent stability can be obtained with this circuit by using negative-temperature coefficient capacitors at C3 through C7 inclusive. The capacitors compensate for the positive drift of the core at L1 (a). The effective capacitance in parallel with L1 is demonstrated in (b).

Drift is not a primary consideration when the equipment remains in a relatively stable temperature environment, where room temperature is fairly constant, and where the ambient temperature within the equipment case undergoes minor changes. But when heat from tubes, power transistors, and transformers is present in the region of the oscillator, long-term drift will become manifest for certain.

A partial solution to the drift problem is to select a slug-tuned inductor (L1 of Fig. 2-26) which has a low-μ core. Also, the coil should be set for the desired inductance with the least amount of core entry into the coil. This condition will not yield the maximum resonator Q, but, as shown in Table 2-4, the Q degradation is not severe. The principal virtue of high Q in an oscillator tank is reduced noise bandwidth at the oscillator output. This is especially important to achieve when the oscillator is used as the LO-injection source for a receiver mixer.

Apart from changes in ambient temperature in the area of the oscillator, long-term drift is likely to result from RF circulating current in the coil and the related capacitors. For this reason it is wise to use a physically large coil, comparatively speaking, which has a fairly large core and a wire gauge (Litz or enameled copper) that can pass the current without significant heating. The more capacitors that are used in parallel at critical RF points (C1 through C6 of Fig. 2-26a) to obtain a specified capacitance, the lower the drift from RF heating. This is because the RF currents are distributed through several capacitors rather than a few, thereby reducing the internal heating of any one capacitor.

In the proven, high-stability oscillator of Fig. 2-26a, the drift from a cold start to a period 3 hours later was so slight that it could not be measured accurately. The maximum shift, as observed by a frequency counter, was less than 1 Hz in a stable temperature environment of 25 °C.

L1 is housed in an aluminum shield, the walls of which should be one coil diameter or greater away from the outer surface of the coil form. Closer spacings degrade the resonator Q and complicate the mechanical stability of the oscillator. An aluminum or brass enclosure will lower the coil inductance somewhat, requiring that the slug be moved farther into the coil winding. Iron types of shields have the opposite effect, causing an increase in inductance. The coil shield is useful in keeping unwanted RF energy from other parts of the circuit (composite equipment) away from the oscillator tank. Also, the shield enclosure helps to maintain a stable temperature when short-term changes in ambient temperature occur.

Polystyrene capacitors are suitable for use at C4, C5, and C6. They exhibit a negative temperature coefficient which compensates for the positive coefficient of the coil magnetic core. NP0 ceramic capacitors are suitable also at those circuit points. Silver–mica capacitors are not recommended in the interest of optimum stability, unless they have known drift characteristics.

The series-tuned Colpitts format was chosen to permit using higher values of inductance at L1 than would be possible in a parallel-tuned arrangement. The major advantage of the high L is that circuit-connecting leads and printed-circuit-board conductors become a smaller part of the overall L than is true when parallel tuning is used. This aids stability and helps to prevent Q degradation. This consideration is especially meaningful at frequencies above 5 MHz.

Oscillator short-term stability is aided by D1, a gate-clamping diode (1N914 or equivalent). The diode limits the positive peaks of the sine wave, in turn leveling the peak transconductance of the FET. This serves as a bias stabilizer and minimizes changes in junction capacitance. The latter, if allowed to run rampant, would generate considerable harmonic energy. D2 stabilizes the drain supply voltage at 8.2.

The derivation of minimum and maximum capacitance range for the oscillator is taken from the equation in Fig. 2-26b. The L and C components have been arranged to illustrate how C_{eff} is determined in this type of oscillator. The effective tuning-capacitance range is approximately 55 to 214 pF with the arrangement shown. This circuit can be extrapolated to frequencies up to 10 MHz by using the reactance values of the capacitors and inductors shown.

An example of high-Q slug-tuned inductors used in a passive bandpass circuit is given in Fig. 2-27. In effect, we are looking at a three-pole Cohn type of tunable filter which might be used to obtain a high order of receiver front-end selectivity. Although the frequency specified is for the standard AM broadcast band, tunable filters of this type can be used at any frequency where lumped constants are acceptable—150 MHz and lower. The important thought to keep in mind is that the loaded bandwidth (BW_L) doubles each time the operating frequency is raised one octave. For this reason a three-pole filter of the type shown is more useful at the lower end of the frequency spectrum.

For the frequency specified in Fig. 2-27, carbonyl E core material is chosen. L2 and L5 comprise individual resonators for the filter. L3 and L4 are used in combination to provide the third resonator. These two inductors have twice the nominal inductance of the end coils so that the net value of the coils in parallel equals that of the individual end sections of the filter—480 and 240 μH, respectively. This scheme enables the designer to use a three-section variable capacitor for tuning, rather than a four-gang capacitor, which would be required if four individual resonators were contained in the filter.

L1 and L6 are adjusted for the desired input and output coupling. L7 and L8 are used as bottom-coupling inductors between the resonators. Small toroidal inductors are suggested for use at L7 and L8. The coupling factors of the filter are chosen to provide the desired BW_L, consistent with

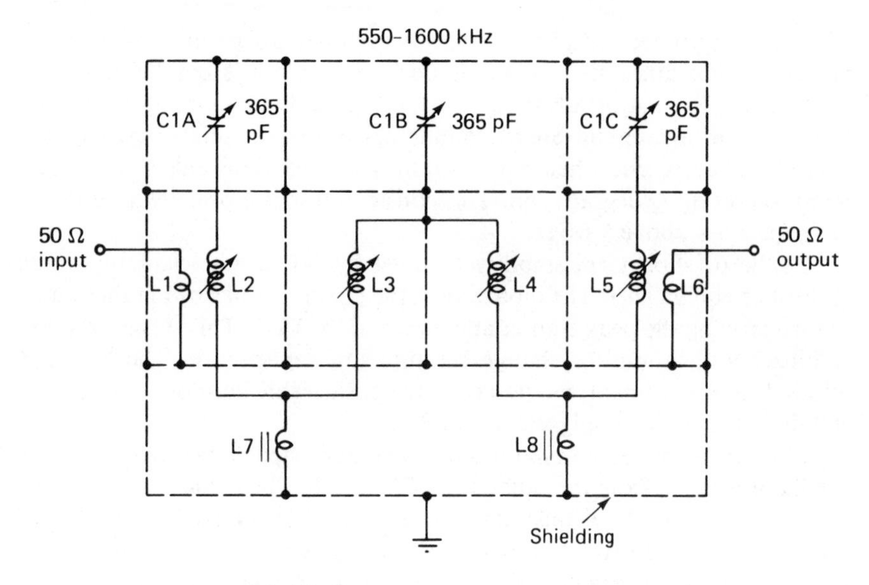

L1, L6: See text
L2, L5: Carbonyl E core, 240 μH nom.
L3, L4: Carbonyl E core, 480 μH nom.
L7, L8: ≈5 μH (see text)

Figure 2-27 A high degree of receiver front-end selectivity can be achieved by utilizing a three-section tunable Cohn filter. High-*Q* slug-tuned inductors are used at L2 through L5. Since L3 and L4 are effectively in parallel, their inductances are twice that of L2 and L5, to permit tracking by means of C1.

the insertion loss that can be tolerated in terms of noise figure and receiver sensitivity. An insertion loss of 4 to 6 dB is a suitable target amount for the frequency specified.

In the test model, the author employed J.W. Miller 43A224CBI inductors at L2 and L5. The inductance range is 138 to 275 μH, with a Q_u of 128 with the core set for 138 μH. Inductors L3 and L4 are J.W. Miller 43474CBI parts, with a range of 240 to 580 μH. Q_u at 240 μH is 135. The Q_u of both coil types was measured at 790 kHz.

The dashed lines in Fig. 2-27 indicate extensive shielding of the filter from outside influences. Each resonator is shielded from the others. C1, the three-section tuning capacitor, should be constructed with shield baffles between the rotor sections to minimize leakage from one filter pole to the next.

This circuit was developed primarily to demonstrate just one application for slug-tuned inductors. There is no reason why toroidal or pot-core units could not be substituted for the type of inductors shown. Were this done, however, it would be necessary to include trimmer capacitors at each section of the filter to permit tracking.

2.6 RF Chokes

It is useful to include magnetic-core RF chokes in many designs, particularly when miniaturization is a keynote. Ferrite or powdered-iron choke cores permit large inductance values to be realized in much smaller volume than is true when air-core or dielectric-core chokes are built. Another advantage to the choke with a magnetic core is that considerably fewer turns of wire are needed for a specified inductance, as compared to a choke of the same value which uses no core material. This greatly lowers the dc resistance of the winding and elevates the Q_u. Figure 2-28 shows a collection of modern RF chokes which are built in various physical formats. Some RF chokes are available in encapsulated form. Others are built with the windings exposed. The designer decides which style is needed, in accordance with the environment in which the choke will be used (humidity, chemical effects, etc.) Figure 2-29 contains pictorial representations of a solenoidal choke (a), a pi-wound unit (b), and a power type of solenoidal RF choke (c). These illustrations portray only three of the myriad formats that RF chokes take. Some have two or more pi windings, others are encased in phenolic material, and so on.

2.6.1 Choke Applications

RF chokes are not selected in random fashion during the design period. There is a definite and necessary relationship between the choke characteristics and the circuit point at which it is used. If the choke is used as a *choke,* in the true sense of the word, the reactance should be considerably higher at the lowest operating frequency than the impedance of the circuit to which it is connected. A viable rule of thumb is to ensure that the RF choke has an X_L which is four times or greater the circuit impedance. This principle is demonstrated in Fig. 2-30, where a cathode-driven RF power amplifier receives excitation in parallel with a cathode choke, RFC1. The input impedance of the amplifier at full drive is approximately

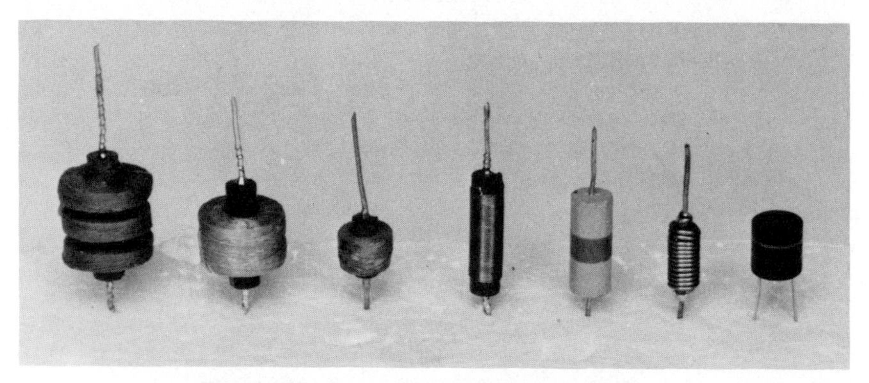

Figure 2-28 Group of magnetic-core RF chokes.

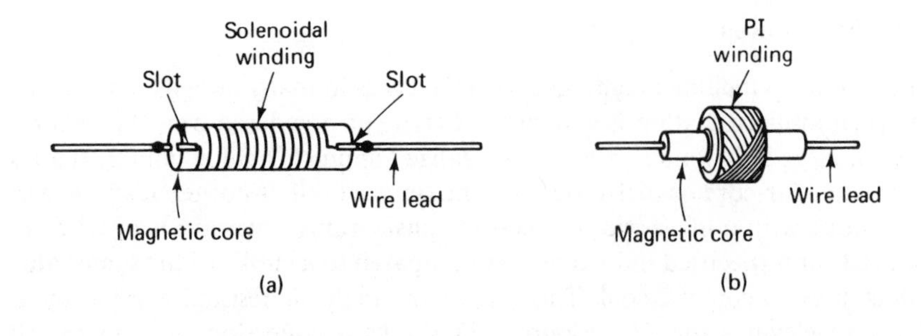

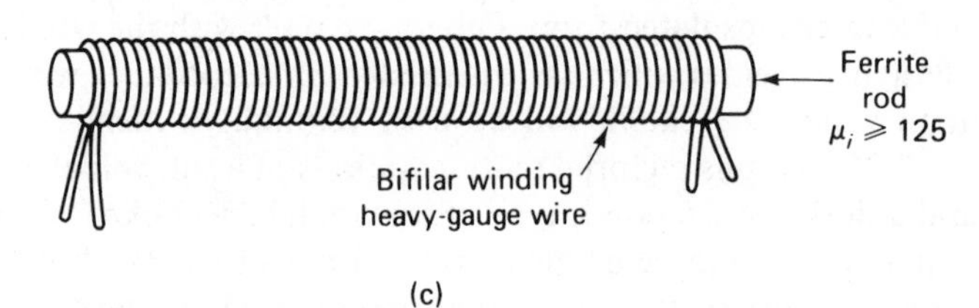

Figure 2-29 A small solenoidal RF choke is seen in (a). A bank-wound RF choke is illustrated in (b). A bifilar-wound power choke is depicted in (c). All three chokes are wound on magnetic-core material.

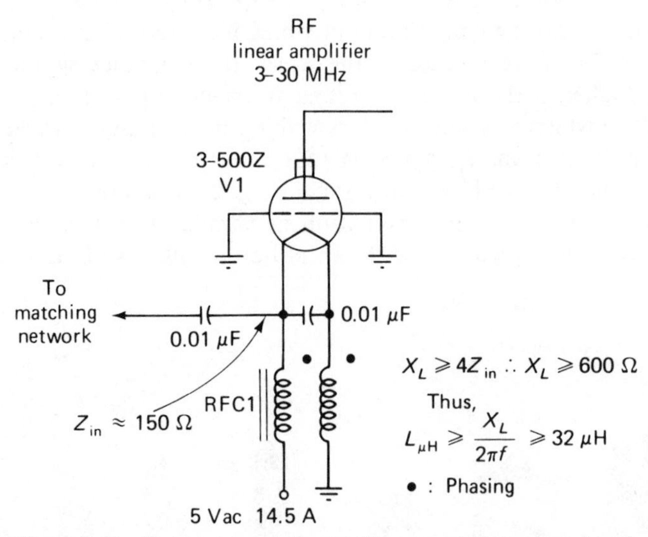

Figure 2-30 Design example for a bifilar-wound filament choke of the type seen in (c) of Fig. 2-29.

150 Ω. Therefore, X_L for RFC1 should be equal to or greater than 600Ω at 3 MHz—the lowest operating frequency. This requirement can be met by using a ferrite-rod core that is 0.5 × 7.5 in., with a μ_i of 125. The required inductance for 600 Ω at 3 MHz can be found by

$$L_{\mu H} = \frac{X_L}{2\pi f} = \frac{600}{6.28 \times 3} = 32\ \mu\text{H}$$

where f is in MHz. Once this is known, the number of turns to obtain 32 μH on the core specified can be determined from the A_L factor of Fig. 2-3c (448). Thus,

$$N = 100\sqrt{L_{\mu H} \div A_L} = 100\sqrt{32 \div 448} = 26 \text{ turns}$$

where N is the number of turns desired.

The filament/cathode choke uses a bifilar winding. This can be considered as a single-wire winding in the calculations. The resultant inductance is essentially the same for a single winding versus a bifilar winding if each winding occupies the same amount of space on the core. The Q remains the same also. So, to obtain 32 μH of inductance we will use 26 bifilar turns of enameled copper wire, close-wound, as shown in Fig. 2-29c.

The bifilar winding can be extended to fill 6 in. of the core, in which case the inductance will be roughly 100 μH (45 bifilar turns of No. 14 wire). The core, thus used, will exhibit a Q_u of 150 at 8 MHz. The No. 14 wire will pass the 14.5-A filament current without excessive heating or IR losses. The core material chosen for RFC1 satisfies the B_{max} requirements treated in Chapter 1. The $X_L \geq 4Z_{\text{in}}$ is necessary to prevent power loss through the RF choke. Furthermore, if the X_L is too low, an intolerable mismatch can result between the signal source and the load when RFC1 is placed in shunt with the 50-Ω driving source.

The assembled RF choke can be coated with glyptol varnish to protect it from damage and moisture. Alternatively, the winding can be encased in heat-shrink tubing. Formvar-insulated copper wire is recommended for the bifilar winding on RFC1 of Fig. 2-30. This insulating material is tough, as well as being resistant to damage from chemicals, heat, and oil.

Some additional applications for RF chokes are given schematically in Fig. 2-31. In this example the magnetic-core inductors are used in the base return of a Class C amplifier (RFC1) and in the V_{cc} line to the amplifier, Q1. Once again we shall observe the 4 × Z rule of thumb for X_L. The base-return choke should exhibit a reactance of 12Ω or more. This requires an inductance of at least 0.83 μH. A standard choke value of 1 μH would be a suitable choice.

RFC2, the collector choke of Q1, has an approximate collector impedance of 3.7Ω, based on the standard relationship $Z = V_{cc}^2/2P_o$, where

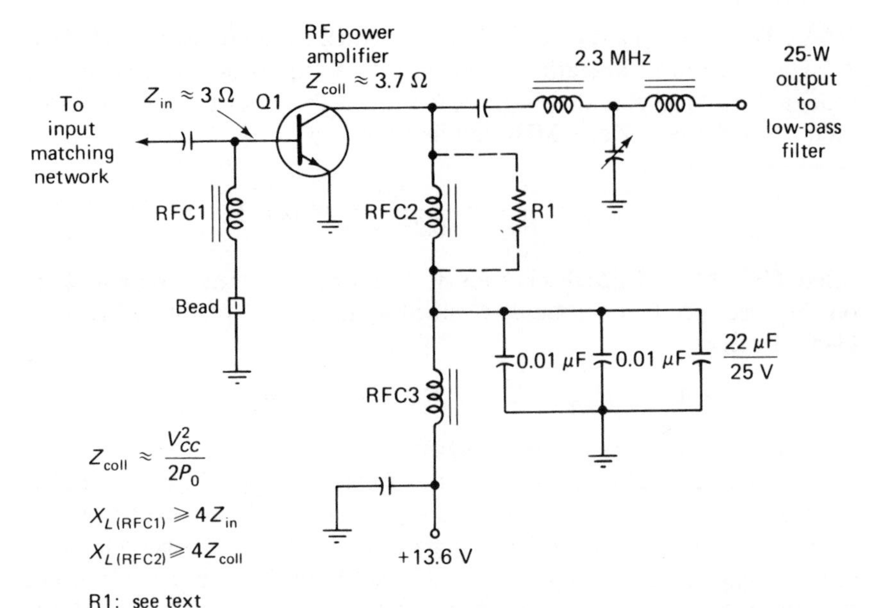

Figure 2-31 Additional uses for magnetic-core RF chokes are given here. RFC1, RFC2, and RFC3 are discussed in the text.

P_0 is the desired output power in watts. Thus, X_L for RFC2 should be 14.8Ω or greater. At four times Z_{coll} this calls for 1.02 μH. Once again, a standard-value choke of 1 μH will suffice.

RFC3 is used as a decoupling choke in the V_{cc} supply line. Since I_{coll} will be approximately 3.67 A, allowing 50% efficiency for the amplifier, the supply-line impedance is roughly equal to 3.7 Ω $(Z = E/I)$, RFC3 should have approximately 1 μH of inductance also.

Power transistors tend to self-oscillate at the operating frequency and elsewhere in the spectrum by virtue of tuned base/tuned collector conditions. The greater the transistor gain, the worse the problem. Typical HF, VHF, and UHF power transistors exhibit gains of 13 dB or more. Stability is, therefore, a design problem if care is not taken with layout and component selection. In theory, the transistor gain increases 6 dB per octave as the frequency is lowered. Because of this phonomenon, self-oscillation at low frequency and VLF is the principal malady to consider. If a power transistor is operated somewhat below its rated saturated power-output rating (the point at which no further output can be obtained with an increase in drive), the beta rises and stability is marginal. This is because as the collector current drops, the beta rises.

When transistor input and output capacitances are similar, and when base and collector chokes are of the same or approximate value, the tuned base/tuned collector syndrome arises. If the RF chokes have high Q, such as they might when ferrite cores are used, the condition is quite apt to happen. Therefore, it is sometimes necessary to swamp the chokes as shown in Fig. 2-31. RFC1 is shown with a single miniature high-μ ferrite bead in series with the grounded end. A bead of 950 μ is slipped over the pigtail of the choke to swamp the choke Q. Laboratory measurements proved that a single bead increased the choke inductance by 1 μH at 2.5 MHz. The choke Q fell from 60 to 10 when the bead was added.

An alternative Q-killer technique is employed at RFC2. R1 is placed in parallel with the choke to lower the Q. The value of resistance should be selected to stabilize the amplifier with the highest ohmic value possible. The lower the resistance of R1, the more RF power will be dissipated in the resistor. Typically, values from 150 to 470 Ω are satisfactory. Bypass capacitors which are effective at various frequencies in the spectrum are used at the junction of RFC2 and RFC3. They are an important part of the decoupling network.

Solenoidal RF chokes need not be used for RFC1 through RFC3. Small toroids can serve as the cores for these chokes. This could enhance stability by means of the self-shielding qualities of toroids.

A wide assortment of blank coil forms with pigtails affixed are manufactured for use as RF-choke cores. They are available in various μ_i values.

REFERENCES

[1] **W.J. Polydoroff**, *High-Frequency Magnetic Materials.* New York: John Wiley & Sons, Inc., 1960.

[2] **Henry Jasik**, *Antenna Engineering Handbook.* New York: McGraw-Hill Book Company, 1961.

3

APPLYING TOROIDAL CORES

The rules governing core selection with respect to physical size, permeability, frequency, and B_{max} were discussed in Chapter 1. In this section we focus our attention on some of the practical applications of toroids in communications circuits. Complete design examples will be given to illustrate core selection and actual utilization of inductors and transformers in typical circuits. Wherever it is possible and practical to do so, rules of thumb will be offered as time-saving shortcuts for the engineer or technician who is engaged in developing prototypes. From these plateaus the proof-build or production-stage circuits can evolve with minimum refinement effort.

3.1 Low-Level Circuits

For the purpose of definition, *low level* in this book refers to circuits in which the power level is less than 1 W. In some engineering circles this term relates to power levels in the mW region. For the most part, the expression is a mathematical one. However, it has been defined here as it will be used by the author.

In communications equipment there are a number of low-level circuits in which inductors or transformers with magnetic cores find application. Prominent among these circuits are RF amplifiers, IF amplifiers, oscillators, mixers, detectors, tuned audio amplifiers, and filters. Some of these circuits operate in narrow-band fashion, while others are classified as broadband in their functions. The narrow-band circuits have reasonably flat responses over a range of hertz or kilohertz, depending on the design. Broadband circuits exhibit responses as great as several GHz, conversely. In general terms, narrow-band circuits are not as complicated to design as are broadband circuits. In either case it is necessary to pay special attention to the core material used, while also considering the operating parameters of the active devices in the related circuit. Our treatment of magnetic cores will be expanded to include other important design features for the composite circuits used as examples.

A narrow-band circuit can be structured to encompass a wide range of frequencies by making it continuously tunable over the required frequency spread. On the other hand, some narrow-band amplifiers are fixed-tuned to operate on a single frequency, such as would be the case with an IF amplifier. Continuously tunable narrow-band circuits are indigenous, primarily, to the front ends of radio receivers and the stages of some multiband radio transmitters, the latter of which employ vacuum tubes.

Magnetic-core material is seldom used in low-level circuits above approximately 150 MHz. The same is not quite as true of high-level circuits, where broadband transmitters contain ferrites in devices called isolators and circulators. Figure 3-1 shows a number of toroidal inductors and transformers which are typical of the kind discussed in this chapter.

3.1.1 RF Amplifiers

A fixed-tuned RF amplifier for use at 2.5 MHz is used as a design example in this section. Figure 3-2 contains the circuit and some pertinent data that apply to it.

Let us assume that Q1 has been chosen for low noise and full gain capability at 2.5 MHz. A suitable rule of thumb for ensuring adequate gain at f_0 is to choose a transistor which has an f_T rating at least 10 times f_0. This calls for an f_T of 25 MHz minimum when selecting Q1. The symbol f_T specifies the frequency at which the small-signal forward-current ratio is 1

Figure 3-1 Various toroidal inductors and transformers.

RF amplifier

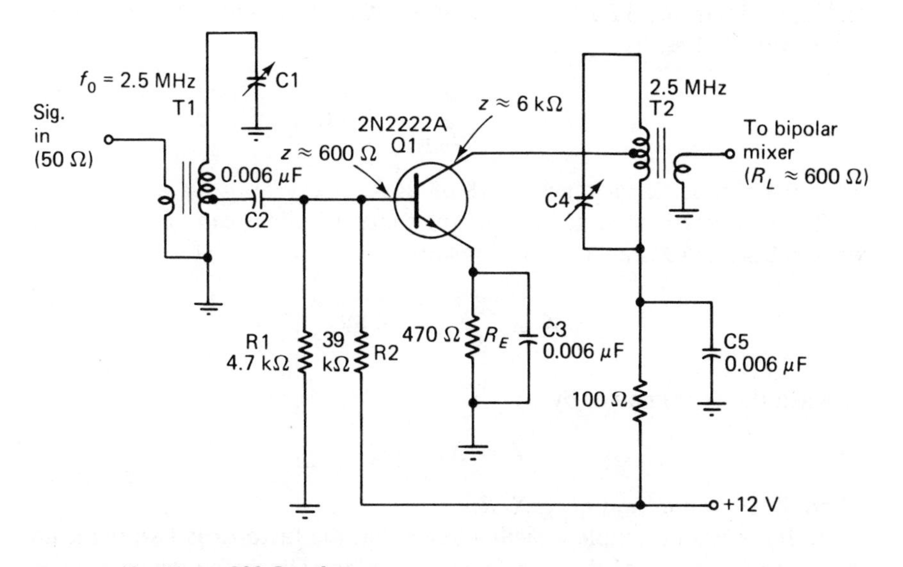

$X_{C1}, X_{C4} \approx 200\ \Omega$ at f_0

$X_{C2}, X_{C3}, X_{C5} \leqslant 10\ \Omega$ at f_0

C1, C4: 318 pF nom.

$N(\text{T1}) = \sqrt{\dfrac{Z_{sec}}{Z_{pri}}}$ $\quad N(\text{T2}) = \sqrt{\dfrac{Z_{pri}}{Z_{sec}}}$

$X_L = X_C$ at f_0

$\therefore X_L$ T1 (sec), T2 (pri) = 200 = 12.7 μH

I_C = 1.5 mA

V_E = 0.7

$R_E = V_{EB}$ (≈ 0.7 silicon)

R1 ≈ $10R_E$

R2: Select to obtain $2V_{EB}$ (1.4 V) with R1-R2 pair

f_T (Q1) $\geqslant 10f_0$

Figure 3-2 Design example for a small-signal RF amplifier in which two toroidal transformers (T1 and T2) are used. Design progression is discussed in the text.

(unity) in the common-emitter state. This is known as the *gain-bandwidth product.* The symbol for small-current forward-current ratio is h_{fe}. The h_{fe} of Q1 must be equal to or greater than the voltage gain required of the amplifier if

$$\frac{\text{N1}}{\text{N2}} = 1.$$

Thus, if we desired a voltage gain of 10 for the circuit of Fig. 3-2, the Q1 h_{fe} should be no less than 10 if the primary and secondary windings of T2 have an impedance ratio of 1. If, with the step-down ratio shown in the diagram, the circuit must provide a voltage gain of 10 at the output port, the h_{fe} will need to be proportionately higher than 10.

An arbitrary X_c is suggested for C1 and C4. It will be 200 Ω, which calls for an X_L of the same ohmic value at resonance. Thus, the secondary of T1 and the primary of T2 have an X_L of 200 Ω. The required inductance of those two windings is

$$L_{\mu H} = \frac{X_L}{2\pi f_{mHz}} = 12.7$$

when the parallel capacitance is 318 pF at f_0.

The characteristic impedance of the tuned windings can be determined when the series resistance, R, is known:

$$Z = \frac{X_L^2}{R} \quad \text{ohms}$$

or when the Q is known, by

$$Z = X_L \times Q$$

where Z is in ohms and $Q = X/R$.

In this design example we will assume that the factor R is 1 Ω, made up by the ac resistance of the transformer winding. This being the case, $Q = X_L/R$ or 200. From the equation for impedance we note that $Z = 200^2/1 = 40{,}000\ \Omega$. Similarly, when X_L and Q are known, $Z = 200 \times 200 = 40{,}000\ \Omega$

Starting with T1 of Fig. 3-2, the turns ratio, primary to secondary, is

$$N = \sqrt{Z_{sec}/Z_{pri}} = \sqrt{40{,}000/50} = 28 : 1$$

The tap point on the secondary winding can be obtained from

$$N = \sqrt{40{,}000/600} = 8 : 1$$

as taken from the basic equation for N.

In making a core selection, we find one that is rated for optimum Q at 2.5 MHz, the f_0 in Fig. 3-2. Carbonyl E will be chosen because of its 1- to 30-MHz recommendation. It has a μ_i of 10, which will permit an ample number of coil turns to obtain a practical realization of the turns ratios determined earlier. A 0.5-in.-diameter core will be large enough to operate without saturating in this type of circuit, assuming that abnormally large signals are not present at the input to T1. An Amidon Associates T50-2 powdered-iron core is suitable (see Appendix E for available Amidon cores).

The A_L factor for a T50-2 core is 50, so to learn how many turns are needed to obtain an inductance of 12.7 μH, we apply

$$n = 100\sqrt{L_{\mu H}/A_L} = 100\sqrt{12.7/50} = 50 \text{ turns}$$

where n is the number of unknown turns and A_L is taken from the manufacturer's literature, or determined as outlined in Chapter 1.

The number of primary turns for T1 is obtained by n/N (50/28), which calls for a primary consisting of 1.78 turns. Since this is an impractical number, a two-turn primary is used. The core size versus wire gauge table indicates that 50 turns of No. 26 enameled wire will fit on a T50 core. To ease the winding job somewhat, No. 28 wire will be used, thereby allowing up to 64 turns to be applied to the core.

The tap point on T1 (secondary) is found by n/N (50/8), which calls for the tap being placed 6 turns above the grounded end of the winding. The calculations for T2 are performed in an identical manner, taking into account the different impedance levels. It is worth observing that irrespective of the computed impedance of the secondary winding of T1 (40,000 Ω), the resultant turns ratio between the primary and the tap point to ground on the secondary, satisfy the 50- to 600-Ω condition. This illustrates the relative unimportance of knowing the Q of the resonator (provided that it is high) or the ac resistance, R, provided that the large winding of the transformer has considerably more turns than the low-impedance winding and tap. Details for biasing Q1 of Fig. 3-2 are beyond the intent of this volume, but data are given in the diagram as guides in approximation. A detailed discussion of amplifiers and biasing methods is contained in a book by Lenk [1].

3.1.2 Capacitive-Divider Matching

It is not always necessary to use the transformer arrangement of Fig. 3-2 for matching the source to the load. The approach seen in Fig. 3-3 simplifies the winding process for toroidal inductors by eliminating the link winding. In this example we find a capacitive divider at the input and output ports of the RF amplifier. Employment of an FET at Q1 eliminates the need for a tap point on L1 and L2, since the gate 1 impedance is in excess of 1 MΩ and the drain impedance with 1.2 mA of current is 10,000 Ω. However, if the input and output impedances of the active device were less than 10,000 Ω, suitable tap points could be placed on L1 and L2 while still using the capacitive dividers.

In order for the scheme of Fig. 3-3 to be applied, R_{in} must be less than R_L. Furthermore, Q_L has to exceed 14.1, as shown by the equation in the design example. This means that the BW_L of the two resonators cannot be of such magnitude as to require a Q_L less than 14.1.

For our design exercise we will specify a BW_L of 500 kHz. This satisfies the Q_L restriction with a resultant value of 20. C_T consists of the nominal value of C1, as computed, in series with the calculated value of C2. Step 4 of Fig. 3-3 provides the required C_T to meet our bandwidth and impedance-transformation parameters. Since X_c and X_L are equal at f_0, and because $X_c = 250$, the required inductance for L1 and L2 is 4 μH. From step 5 we learn that $C1_{nom} = 68.4$ pF and C2 = 833 pF, as defined by DeMaw [2].

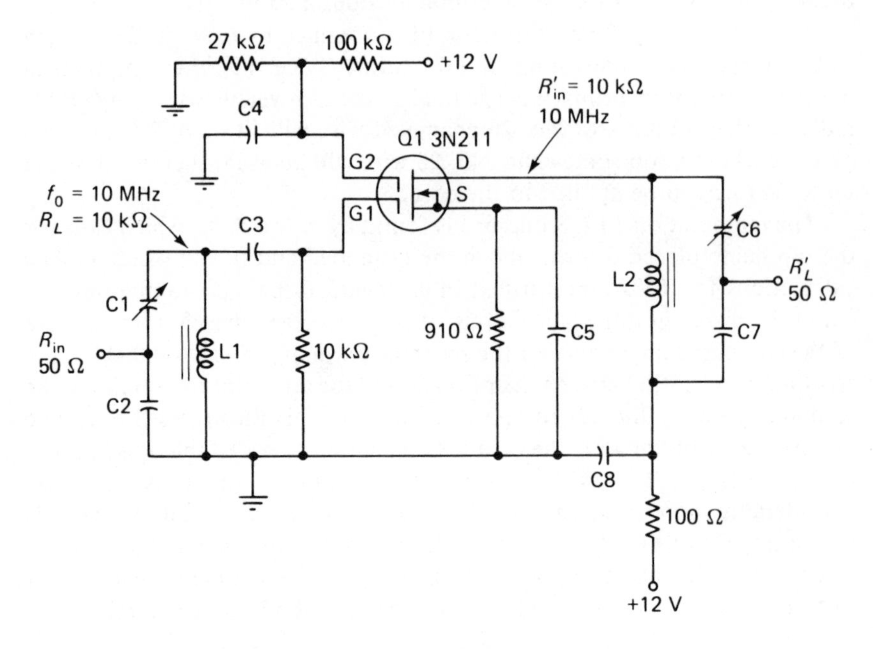

X_C (C3, C4, C5, C8) $\leqslant 10\ \Omega \geqslant 0.0015\ \mu$F

$R_{in} < R_L$ and $R'_L < R'_{in}$

$$Q_L > \sqrt{\frac{R_L}{R_{in}} - 1} \quad \text{and} \quad Q'_L > \sqrt{\frac{R'_{in}}{R'_L} - 1} \quad \therefore\ Q_L > 14.1$$

1: Select BW_L (0.5 MHz)

$$2:\ Q_L, Q'_L = \frac{f_0}{f_2 - f_1} = \frac{10\text{ MHz}}{0.5\text{ MHz}} = 20$$

3: $C_T = C1_{nom} + C2$ and $C'_T = C6_{nom} + C7$

$$4:\ C_T = \left[\frac{Q_L}{2\pi f_0 (R_L/2)}\right] \times 10^6 = 63.6\text{ pF} = X_C\ 250\ \therefore\ L_{\mu H} = 4\ \mu\text{H}$$

$$5:\ \frac{C2}{C1_{nom}} \approx \sqrt{\frac{R_L}{R_{in}} - 1} = 13.1$$

$\therefore\ C1_{nom} = 68.4$ pF, $C2 = 833$ pF, $C6_{nom} = 68.4$ pF
and $C7 = 833$ pF

Figure 3-3 Design data for using capacitive dividers to effect an input and output match when narrow-band toroidal inductors are used without taps or link windings.

From a practical viewpoint, selection of the toroid core is less difficult in this example than was the case related to Fig. 3-2: It is not important how few turns are used in the winding when the smaller primary is not required. The core material should, of course, be suitable for the chosen f_0. A powdered-iron core with an SF mix will result in an optimum Q_u (Q_u should always be considerably greater than Q_L). An Amidon Associates T50-6 core is used. It is an SF type, has a μ_i of 8 and is recommended for use between 2 and 50 MHz. The core diameter is 0.5 in. The A_L factor for this toroid core is 40. Therefore, the required number of turns for L1 and L2 of Fig. 3-3 is 31.6, using the equation from Sec. 3.1.1. A winding of 31 or 32 turns will be acceptable.

Mica compression trimmers are suitable for use at C1 and C6 of Fig. 3-3. Silver–mica capacitors can be used at C2 and C7. Both types will ensure adequate capacitor Q to prevent Q degradation of the two resonators. C1 and C6 are specified as variable capacitors to allow peaking at precisely 10 MHz. This makes it possible to use the nearest standard capacitor value at C2 and C7, which in this design would be 820 pF. Fixed-value capacitors could be placed at all four points if L1 and L2 were replaced by slug-tuned inductors of the type treated in Chapter 2. However, the toroids are a better choice in the interest of circuit stability, owing to their self-shielding characteristics.

3.1.3 Interstage Coupling

Tuned toroidal transformers are excellent as coupling elements between amplifier stages in a receiver or transmitter, as shown in Fig. 3-4. For the purpose of illustration we will assume that the g_{fs} of Q1 and Q2 is 3000 and the drain impedance (Z_d) of Q1 is approximately 4000 Ω. Z_{in} of the source-driven amplifier, Q2, is roughly 333 Ω when g_{fs} is 3000. These characteristics call for a T1 turns ratio of 3.4:1 to ensure maximum transfer of energy from Q1 to Q2. C1 is used to tune the primary of T1 to resonance at f_0. Using an arbitrary X_L and X_c of 200 Ω, the nominal capacitance of C1 at resonance is 454 pF and the inductance of the primary of T1 is 18 μH. In order for C2 and C3 to function effectively as bypass capacitors, X_c should be 10 Ω or less. A 0.01-μF capacitor will suffice as a close standard value. R1 is chosen for the bias required at Q2.

An SF type of powdered-iron core will be suitable for T1 to satisfy the high-Q requirement of the tuned transformer. An Amidon T50-2 core will be satisfactory (A_L = 50), requiring 60 turns of No. 30 enameled wire for the primary winding. The secondary of T1 will contain 18 turns of wire. Although we have specified a type SF core, a Carbonyl C core would be suitable also. It would require fewer turns of wire because of the higher μ (20). The SF core has a μ of 10. If a C-type core was used, we could select an Amidon T50-1, which has an A_L of 100. That being the choice, the primary

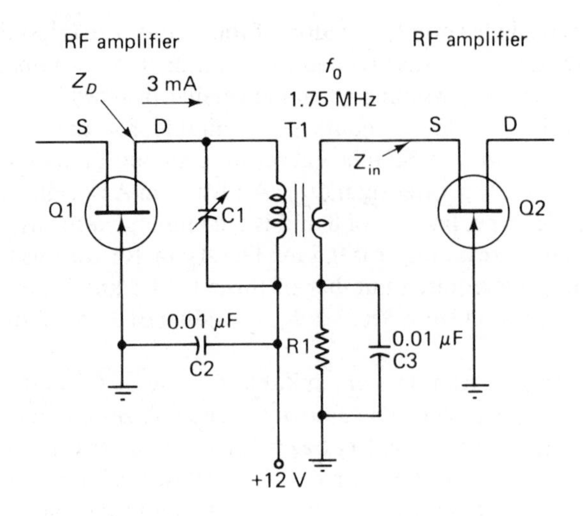

Q1, Q2 (g_{fs}) = 3000

$$Z_D \approx \frac{V_{DS}}{I_{DS}} = \frac{12\text{ V}}{0.003\text{ A}} = 4000\ \Omega$$

$$Z_{in} \approx \frac{10^6}{g_{fs}} \approx 333.0\ \Omega$$

$$N\,(\text{T1}) = \sqrt{\frac{Z_{pri}}{Z_{sec}}} = 3.4$$

X_C (C2, C3) $\leqslant 10\ \Omega = 0.009\ \mu$F min.

$X_{C1} \approx 200\ \Omega$ at $f_0 = 454$ pF nom.

X_L (T1 pri.) $= 200\ \Omega = 18\ \mu$H

Figure 3-4 Toroidal transformer (T1) used for interstage coupling.

of T1 would contain 42 turns of No. 28 wire and the secondary would consist of 12 turns of wire. Either type of core material will yield excellent Q at 1.75 MHz.

A trifilar-wound toroidal transformer is illustrated in Fig. 3-5. In this circuit a mixer, Q1, supplies signal voltage to a half-lattice crystal filter, FL1. T1 is tuned by means of trimmer C1. The transformer phasing is such that Y1 and Y2 are supplied with energy that has a 180° differential. R1 is used to improve the mixer IMD and R2 is chosen to provide the filter termination that ensures minimum passband ripple.

The rule of thumb for X_L and X_c in Fig. 3-4 can be used in this circuit example. Therefore, the windings of T1 will have an inductance of 3.5 μH and C1_{nom} will be 88 pF. An SF type of powdered-iron core will be appropriate for use at 9 MHz. If we employ an Amidon T50-2 toroid core, which has an A_L factor of 50, the transformer will contain 26 turns of wire for each winding. Since the three windings are applied in parallel, the core will require an equivalent of 79 turns. This calls for No. 30 enameled wire. To ease the winding assignment, one could use a larger core of the same material, such as a T68-2. The required number of turns would be 25

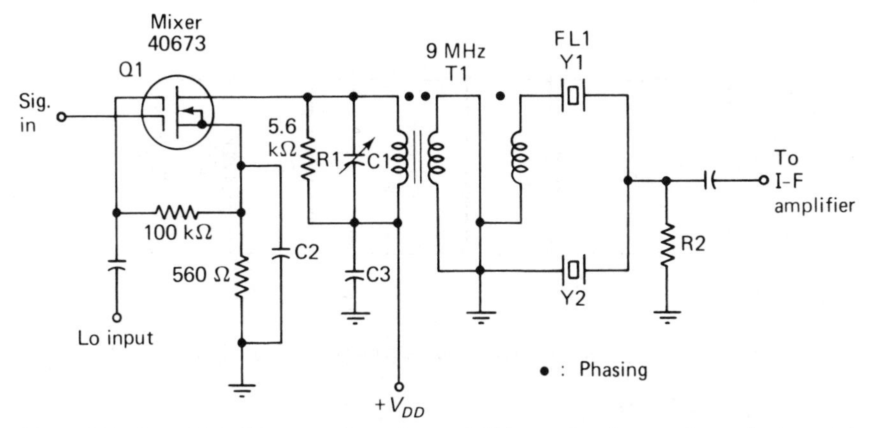

Figure 3-5 A toroidal trifilar-wound transformer (T1) is excellent for coupling a single-ended mixer to a half-lattice filter.

because of the higher A_L (55). Since a 25-turn trifilar winding is approximately equivalent to 75 turns, No. 28 wire would satisfy the requirement. C2 and C3 follow the rule set for Fig. 3-4. The application of bifilar and trifilar windings can be made easier by first twisting the wires approximately eight times per inch. Identification of the windings can be greatly simplified by using three different colors of enamel insulation. Red, green, and brown are the three most common colors available. The wire sizes specified in the foregoing text are those which will fit on the cores indicated for T1 of Figs. 3-4 and 3-5.

3.1.4 Toroids in Filters

One of the most appropriate circuit applications for magnetic-core inductors is in filters. Toroids or pot cores are the choices of most designers when the operating frequency is at high frequency and lower. The advantage in using either type of core is the high Q attainable and the self-shielding properties of toroids and pot cores: Isolation between the filter poles is essential to proper performance.

A design example for a bandpass filter with a 600-Ω characteristic impedance is given in Fig. 3-6. It is based on the image-parameter concept. To ensure a center frequency (f_0) of 250 kHz, L1 requires 77.18 μH of inductance and C2 must have 4336 pF of capacitance. C1, the top-coupling capacitor, requires 556 pF of capacitance. C1 and C2 should be made partially variable by placing trimmers in parallel with fixed-value capacitors. This will allow for final optimizing of the filter to compensate for parasitic inductances and capacitances which may exist. High-Q capacitors of temperature-stable characteristics should be used. Polystyrene, silver-mica, or glass capacitors are suggested.

The toroid cores for use at L1 are chosen for optimum Q at f_0. Carbonyl

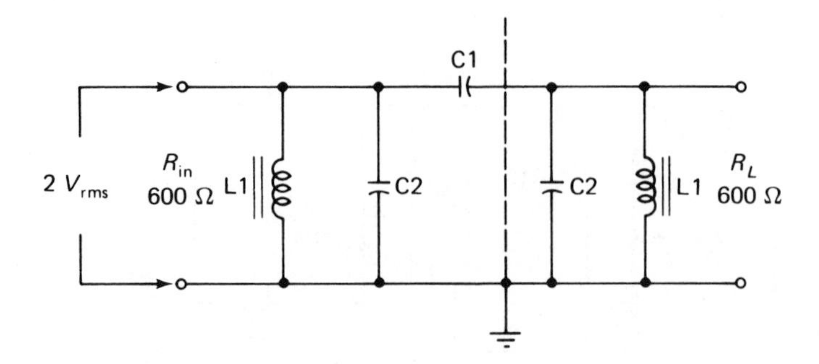

$R = 600\Omega$ $f_0 = 0.250$ MHz $BW_L = 0.05$ MHz $f_2 = 0.275$ MHz

$f_1 = 0.225$ MHz

$$1 \text{ -- } Q_L = \frac{f_0}{f_2 - f_1} = 5$$

$$2 \text{ -- } C1 = \frac{7.96\,(f_1 + f_2)\,10^4}{f_1\, f_2\, R} = \frac{20{,}660}{37{,}125} = 556 \text{ pF}$$

$$3 \text{ -- } C2 = \frac{\left[\dfrac{3.18 f_1\,(10^5)}{f_2(f_2 - f_1)R}\right]}{2} = \frac{\left[\dfrac{71{,}550}{8.25}\right]}{2} = 4336 \text{ pF}$$

$$4 \text{ -- } L1 = 2\left[\frac{.0796\,(f_2 - f_1)\,R}{f_1\;f_2}\right] = 2\left[\frac{2.388}{.0618}\right] = 77.18\ \mu\text{H}$$

5 -- Choose a toroid core (Amidon T68-15). $A_e = 0.196$ cm^2, $A_L = 180$, OD = 0.68 in., $\mu_i = 25$ and $B_{max} = 22$ kilogauss.

$$6 \text{ -- } N = 100\sqrt{L_{\mu H}/A_L} = 100\sqrt{77.18/180} = 65 \text{ turns}$$

$$7 \text{ -- } B_{op} = \frac{E_{rms} \times 10^8}{4.44\, A_e\, N\, (f_{mhz} \times 10^6)} = \frac{2 \times 10^8}{14{,}141{,}400} = 14.1 \text{ gausses}$$

Figure 3-6 Design progression for using toroid cores in a bandpass type of two-section filter.

GS-6 is suitable. It has a μ of 25 and is recommended for use from 0.1 to 2 MHz. An Amidon T68-15 will be satisfactory for use at L1. The equations of Fig. 3-6 demonstrate clearly that B_{op} is only a fraction of the B_{max} rating of the core. Therefore, saturation will not occur.

Another design example is offered in Fig. 3-7. The circuit shows how a toroidal inductor can be used effectively in a tuned audio amplifier to obtain a peak response at 700 Hz. The value of C1 is chosen so that the com-

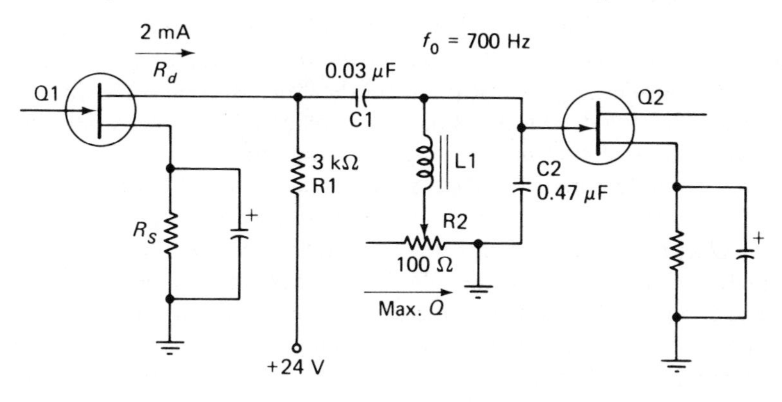

$R_p \approx R_d \| \mathrm{R1} + X_{\mathrm{C1}}$

$X_{\mathrm{C1}} = 7583\ \Omega$

$R_{p\,(\mathrm{eff})} \approx R_d \| \mathrm{R1} + X_{\mathrm{C1}} = 9833\ \Omega$

$X_{\mathrm{C2}}, X_{\mathrm{L1}} = 483$

1. Select R_s for I_d of 2 mA

2. $Q = \dfrac{R_p}{X_{\mathrm{L}}} = \dfrac{9833}{483} = 20.3 \qquad Q = \dfrac{X_{\mathrm{L1}}}{\mathrm{R2}}$

3. $BW_L = \dfrac{f_0}{Q_L} = \dfrac{700\ \mathrm{Hz}}{20.3} = 34.5$ Hz with R2 at 0 Ω

4. $BW_L = 143$ Hz with R2 at 100 Ω

5. Choose a toroid core for L1 (110 mH). Amidon FT-82-72 ferrite, $\mu_i = 2000$, $A_L = 1172$, OD = 0.8 in.

6. $N = 1000\sqrt{L_{\mathrm{mH}}/A_L} = 1000\sqrt{110/1172} = 306$ turns

Figure 3-7 A high-Q toroidal inductor is employed in this design example of a tuned audio amplifier. R2 serves as a Q (bandwidth) control.

bination of C1 and R1 will be high enough in value to prevent excessive loading of the resonator, L1/C2. A value of approximately 9833 Ω results from the values indicated. Once this information is established, we can choose an arbitrary value for X_{C2} and X_{L1}, which for the purpose of this demonstration is 483 Ω. The design procedure can now progress as shown in steps 1 through 6 of Fig. 3-7. The bandwidth of the resonator can be varied from roughly 35 Hz to 143 Hz by means of R2. As resistance is added at R2, the resonator Q drops and the bandwidth increases.

In order to obtain a practical realization of the inductor, L1, a high-μ ferrite core is selected. The tuned circuit will be resonant at 700 Hz when C2 is 0.47 μF and L1 has 110 mH of inductance. An Amidon FT-82-72 toroid core meets our requirement and will contain 306 turns of wire to provide the inductance specified. The skirt selectivity of the response curve can be enhanced considerably by cascading another filter section after Q2. A dual 100-Ω control could then be employed in place of R2. This would permit bandwidth control over both resonators, simultaneously.

Toroidal inductors are well suited to the harmonic filter depicted in Fig. 3-8. In this example cascaded pi networks are used as a half-wave filter to reduce the harmonic currents of a 5-MHz local oscillator. Since the characteristic impedance of the bilateral filter is 50Ω, and because a Q_L of 1 is specified, X_L for the two inductors will be 50. X_c for C1 and C3 is also 50, and X_c for C2 is 25. Steps 1 through 6 illustrate the simple progression for designing the filter to have an f_{c_0} (−3 dB) of 6 MHz. In most practical situations where this type of filter is used, the nearest standard capacitor value can be employed.

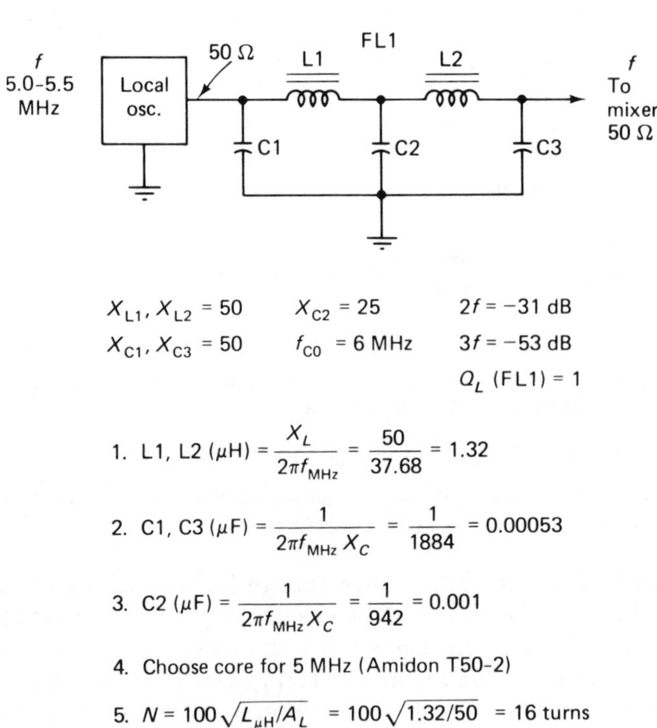

Figure 3-8 Toroids are well suited for use in harmonic filters because of their self-shielding properties and compactness. Here L1 and L2 are used in a half-wave low-pass filter.

A five-pole Chebyshev filter with a low-pass response is presented in Fig. 3-9. In this example there are three toroidal inductors employed, L1, L2, and L3. A maximum RF power level of 1000 W is specified, which results in an rms voltage level of 224 across 50 Ω. With the inductive reactances specified the end inductors require 10.1 μH of inductance and the center coil has an inductance of 14.28 μH. The f_{co} (−3 dB) is 1675 kHz. C1 and C2 compute to the standard value of 0.002 μF. Silver-mica capacitors with a rating of 600 V will be suitable at C1 and C2, provided that an SWR of 1 prevails in the transmission line.

Carbonyl C material is selected for the toroids. It will provide adequate Q at the operating frequency of the filter. The μ_i of the material is 20. An Amidon T106-1 core is used. It is 1.06 in. wide and 0.437 in. thick. The ID is 0.56 in. The maximum acceptable flux density of the material chosen is 12,000 G. Design steps 3 and 4 of Fig. 3-9 yield a B_{op} of 235 G for L1 and L2, with a saturation density of 194 G for L2. (B_{op} is the operational flux

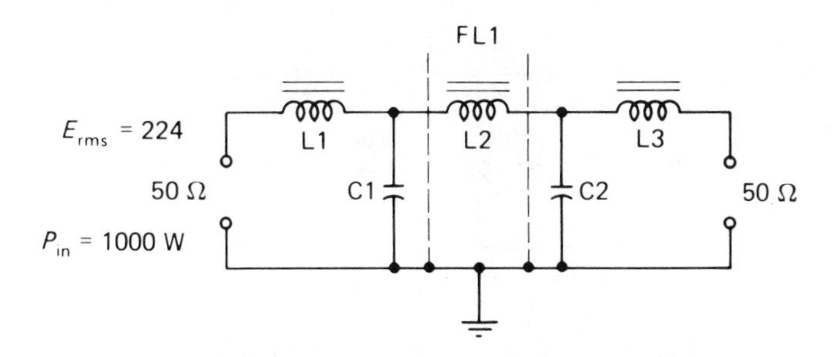

5-pole Chebyshev

$X_{L1}, X_{L3} = 106.8$	$X_{L2} = 150.3$
$X_{C1}, X_{C2} = 45.8$	$2f = -45$ dB
$f_0 = 1600$ kHz	$3f = -65$ dB
$f_{co} = 1675$ kHz	$4f = -78$ dB

L1, L3 = 10.1 μH L2 = 14.28 μH C1, C2 = 0.002 μF

1. Choose a toroid core (Amidon T106-1, $A_L = 280$, 12,000 G, $A_e = 0.706\ \text{cm}^2$)

2. L1, L3 – 19 turns No. 16 wire. L2 = 23 turns No. 16 wire

3. $B_{op}\ (\text{L1, L3}) = \dfrac{224 \times 10^8}{4.44 \times 0.706 \times 19 \times 1.6 \times 10^6} = 235\ \text{G}$

4. $B_{op}\ (\text{L2}) = \dfrac{224 \times 10^8}{4.44 \times 0.706 \times 23 \times 1.6 \times 10^6} = 194\ \text{G}$

Figure 3-9 Core selection must be based on the B_{op} of the circuit. L1, L2, and L3 are chosen accordingly in this design example.

density and is a more definitive term than B_{max}, which was introduced earlier in the book. Both terms find common usage in the industry, and they refer to the same condition.)

Both values are well below the level at which saturation takes place. Some heating of the powdered-iron cores will occur. Therefore, it is important that ample ventilation be included in the design.

Some RF engineers prefer to use E_{peak} rather than E_{rms} for the B_{op} equation. This will allow additional leeway in choosing a core that will operate in the linear portion of its curve. However, when B_{op} is considerably below B_{sat} for a given core, as is the case in Fig. 3-9, the E_{rms} factor is satisfactory.

An example of a high-pass filter that contains toroidal inductors is given in Fig. 3-10a. A filter of this type would be used at the 50-Ω input of a marine-band receiver to prevent broadcast-band signals from overloading the marine receiver front end. This design is based on the image-parameter concept.

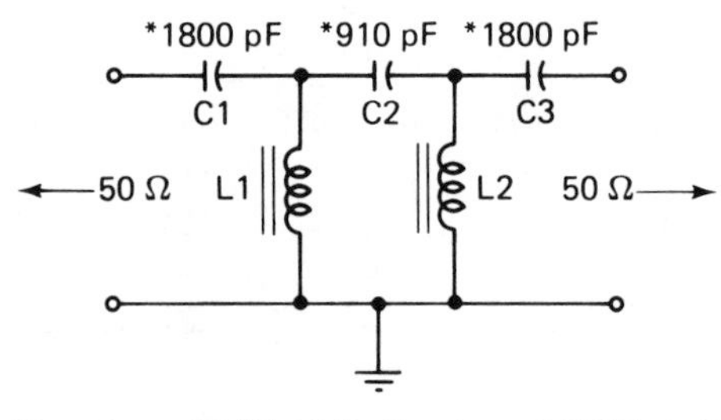

X_{L1}, X_{L2} = 24.55 = 2.3 μH f_{co} = 1.7 MHz (−3 dB)

X_{C1}, X_{C3} = 50 = 1873 pF

X_{C2} = 100 = 936 pF

L1, L2 – 21 turns No. 22 wire on SF core (Amidon T50-2)

* – Nearest standard value

(a)

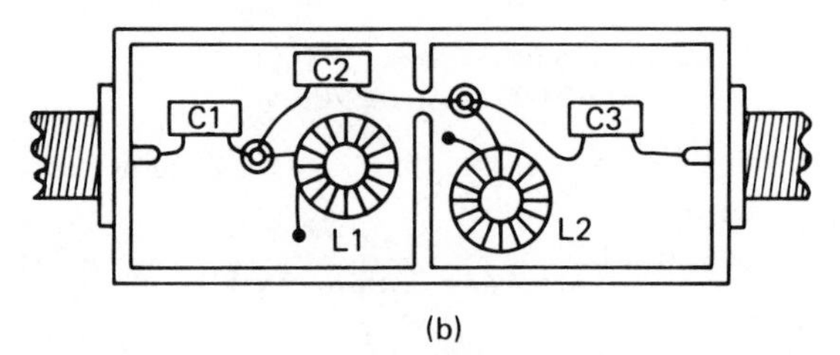

(b)

Figure 3-10 The circuit in (a) uses two toroidal inductors in a high-pass filter. The pictorial view in (b) suggests a method for discouraging unwanted capacitive coupling between the filter sections.

The core material for L1 and L2 is type SF powdered iron. An Amidon T50-2 core is used. To provide the required 2.3 μH of inductance, the cores will be wound with 21 turns of No. 22 enameled wire. C1, C2, and C3 compute to values slightly different than those given in the schematic diagram, so the nearest standard values have been specified. The change in filter performance will be minimal when using the values listed in Fig. 3-10a.

Figure 3-10b illustrates the recommended way to assemble the filter. The same layout technique is suggested for all the filters described in this section. Although toroids have a self-shielding characteristic, it is best to isolate them from one another in shielded compartments. This will prevent unwanted stray coupling between the leads. Capacitive coupling is also discouraged by this method. A die-cast enclosure is suitable for a commercial product line. Double-clad printed-circuit-board material is easy and inexpensive to use for prototypes and laboratory models. The silver–mica capacitors can be supported by their pigtails, as shown.

3.1.5 Toroids in Instruments

There are countless applications for toroids in test instruments. We will examine a few examples here for the purpose of illustration. The first presentation is that of a simple 20-dB bidirectional coupler (Fig. 3-11). The circuit is used as an attenuator of the nondissipative variety. Ports X and Y are for the signal source and a 50-Ω load. A straight length of wire connects X and Y, passing through the core to consitute a single turn of the current transformer, T1. The secondary winding of T1 contains 10 turns. A 0.5-in.-diameter ferrite core is used. It has a μ of 950 (Amidon FT50-43 core), ensuring broadband characteristics throughout the high-frequency spectrum. When port Z is terminated in 50 Ω resistive, this load will be reflected through T1 in accordance with the turns ratio squared. As a result, the core will appear to be a 0.5-Ω resistance in series with the through-line. This condition will cause the source at port X to look into 50.5 Ω. Since the

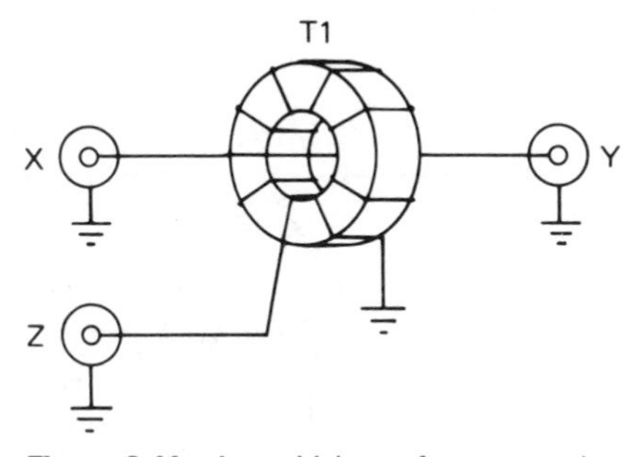

Figure 3-11 A toroidal transformers can be used as shown here to construct a 20-dB bidirectional coupler (see the text).

ratio of the resistances is 50/0.5, or 100, the power delivered at port Z will be −20 dB of that appearing at port Y. The usefulness of this type of attenuator is seen when the signal source is too great in magnitude to use safely with some instrumentation.

An extension of the circuit given in Fig. 3-11 is shown in Fig. 3-12. In the example given, we see an SWR indicator for use in an RF transmission line of 50 Ω impedance. T1 is a current transformer with the equivalent of a one-turn primary. The voltage appearing across the transformer secondary is proportional to the current in the through-line, as is the voltage across R1. The voltage appearing at the junction of C1 and C3 is reduced from that on the through-line by the ratio of the capacitive divider. C1 is adjusted so that the voltage at C1, C3, and D1 is the same as that across R1. When J2 is terminated in a 50-Ω load, the two voltages will be equal and of opposite phase, thereby canceling. This will provide a reading of zero at M1 to indicate a VSWR of 1.

Terminals J1 and J2 are now reversed and C2 is adjusted to provide a null at M1 with J1 terminated in 50 Ω. The bridge will at this juncture be adjusted properly for a 50-Ω characteristic impedance. With the load con-

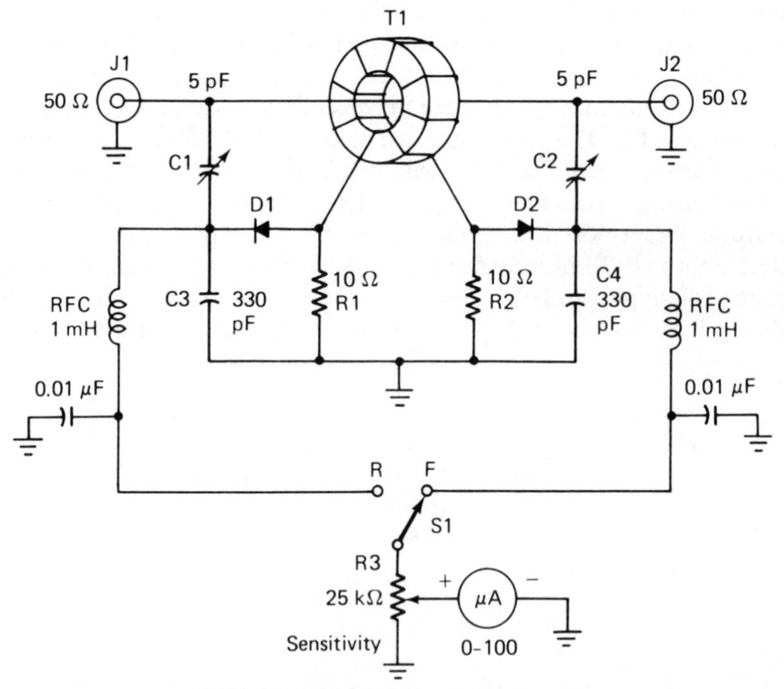

Figure 3-12 The basic configuration of Fig. 3-11 can be used as shown here to provide an SWR indicator or RF power bridge. This circuit is based on the classic design by Warren Bruene of Collins Radio Co.

nected to J2, the source to J1 and the meter switched to read forward power (F), a full-scale reading should be noted at M1 when R3 is adjusted for greater resistance than was present during the adjustment of C1 and C2. Typically, R3 is set for maximum meter sensitivity during the nulling procedure. Thereafter, the control is adjusted for full scale at M1 when maximum power is flowing through the bridge.

The circuit is suitable for relative SWR measurements, as shown in Fig. 3-12. Additional switches and trim pots can be inserted between S1 and the two RF chokes to provide calibrated RF power readings in both the forward and reflected modes. This technique permits actual RF power measurements in both directions. An accurate means of measuring VSWR will result.

The arrangement shown at Fig. 3-12 is patterned after the basic RF power bridge developed by Warren Bruene of Collins Radio in the 1950s. Most commercial RF bridges that are used for power and SWR measurement (using a current transformer) are founded on the Bruene design.

Another useful application of a toroidal transformer is seen at Fig. 3-13. This circuit can be used as a return-loss bridge for measuring impedance. It can be employed also as a 6-dB hybrid combiner when, for example, two signal generators must be connected to a test circuit. An example of such an application can be envisioned where the dynamic range of a receiver must be measured.

When the circuit of Fig. 3-13 is used as a return-loss bridge, the unknown impedance is connected to port X. Initial setup calls for applying power to the bridge while port X is open- or short-circuited. A 50-Ω detector is connected to port Y. Signal power is increased until a full-scale reading is obtained on detector meter. Next, the unknown impedance is attached at port X. With the same level of power applied to the bridge, the meter is observed for scale reading. The ratio of the meter readings, expressed in decibels, is the return loss (dB $= 20 \log E_1/E_2$). The greater the return loss,

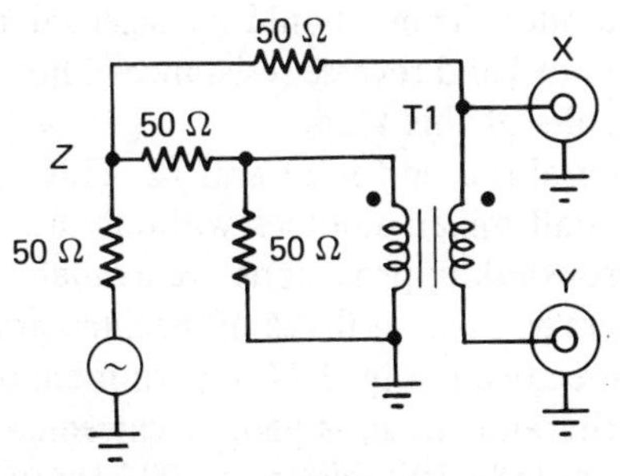

Figure 3-13 A toroidal transformer can be used effectively as the heart of a return-loss bridge for impedance measurements.

the closer the unknown impedance is to 50 Ω. T1 is wound on a ferrite toroid core of 950 Ω (Amidon FT50-43). It contains 10 bifilar turns of No. 24 enameled wire. This basic circuit was described by Hayward in *Solid State Design for the Radio Amateur* [3].

When the circuit of Fig. 3-13 is used as a 6-dB hybrid combiner, a pair of signal sources are connected to ports Z and Y. The equipment under test is attached to port X. If each generator is adjusted to deliver 20 mV to the combiner, an output voltage of 10mV will result, thereby causing a loss of 6 dB. The isolation between the generators is desirable to prevent one signal source from phase-modulating the other. If phase modulation were allowed to take place, the resultant sidebands could render the measurements inaccurate.

3.1.6 Other Low-Level Applications

Broadband passive and active mixers depend upon toroidal transformers in most instances to provide a balanced-to-unbalanced (balun) transformation. The same is true of balanced-modulator circuits of the broadband variety. Figure 3-14 contains details for a basic doubly balanced mixer (DBM) of the four-diode passive type. The same configuration can be used for diode-ring balanced modulators. In the latter case the audio is applied at J3 and the suppressed-carrier RF energy is taken out at J2.

The terminal impedance of the circuit shown in Fig. 3-14 is approximately 50 Ω when one of the ports looks into a 50-Ω load. Some modern balanced mixers of the DBM type used in this illustration are followed at J3 by a diplexer. This can consist of an *LC* network that is terminated in 50 Ω. The f_{co} of the filter is roughly three times the IF. This technique will help to ensure a 50-Ω termination for the mixer at all three frequencies of interest, thereby improving the mixer IMD characteristics.

To realize optimum balance, it is necessary that T1 and T2 of Fig. 3-14 be as symmetrical as possible when installed in the circuit. The same holds true for the four diodes. They should be selected for dynamic balance (closely matched forward and reverse resistances): hot-carrier diodes offer a good match for circuits of this kind.

High-μ core material is used for T1 and T2. This permits the designer to develop relatively small mixer modules without sacrificing low-frequency response. When core-winding provisions are available for very light gauge wire, toroid diameters as small as 0.125 in. become practical. Quarter-inch-diameter cores are specified in Fig. 3-14 to permit hand winding the toroids. The material is ferrite and the μ_i is 950. If extreme care is given to symmetry, this circuit should be suitable from 500 kHz to as high as 500 MHz. Many commercially made DBMs are rated from the low-frequency spectrum to above 1000 MHz.

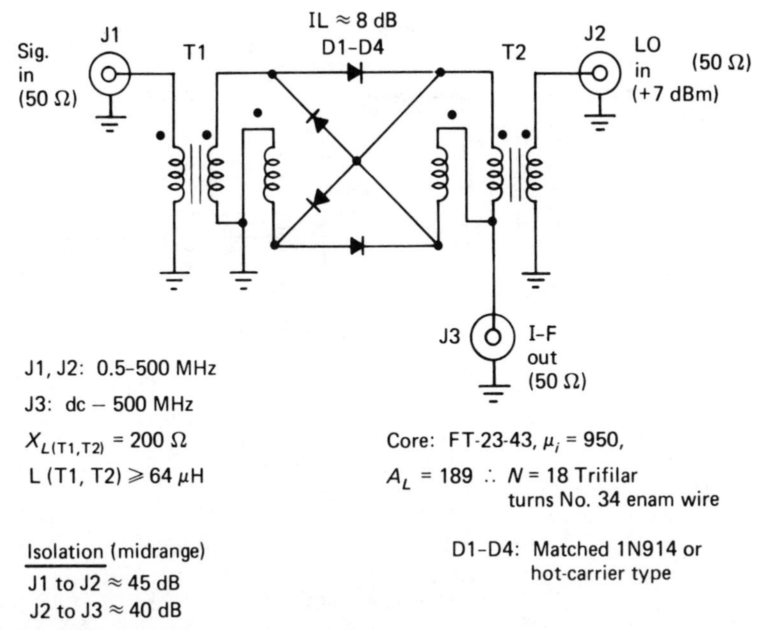

Figure 3-14 Doubly balanced mixers contain broadband baluns (T1 and T2) as seen here. These trifilar-wound transformers help to ensure symmetry, which is vital to balanced mixers.

Port-to-port signal isolation at the spectral midrange of this mixer is approximately 45 dB from J1 to J2. Between J2 and J3 it should approach 40 dB.

X_L for the trifilar windings of T1 and T2 should be roughly four times the circuit impedance at the lowest operating frequency. Therefore, an X_L of 200 Ω will suffice, thus requiring some 64 μH of winding inductance. This calls for 18 turns of No. 34 enameled wire on an Amidon FT-23-43 or equivalent core.

The basic circuit of Fig. 3-14 is useful as an RF switch or RF attenuator when bias is applied to the diodes in the appropriate fashion. In such cases the B_{op} of the transformers must include the dc bias as well as the ac voltage applied to them.

Figure 3-15a demonstrates how a broadband toroidal transformer (T1) can be used in a diode type of frequency doubler. In the same circuit a narrow-band toroidal transformer, T2, is employed. A trifilar winding consisting of 10 turns of No. 28 enameled wire on a 0.375-in.-diameter ferrite toroid (950 μ) will suffice for T1. A powdered-iron toroid core is recommended for use at T2. This will ensure a high Q for the tuned circuit—a necessary step toward "laundering" the $2f$ waveform. If C1 has a nominal capacitance of 40 pF, the secondary of T2 will require 0.7 μH of inductance.

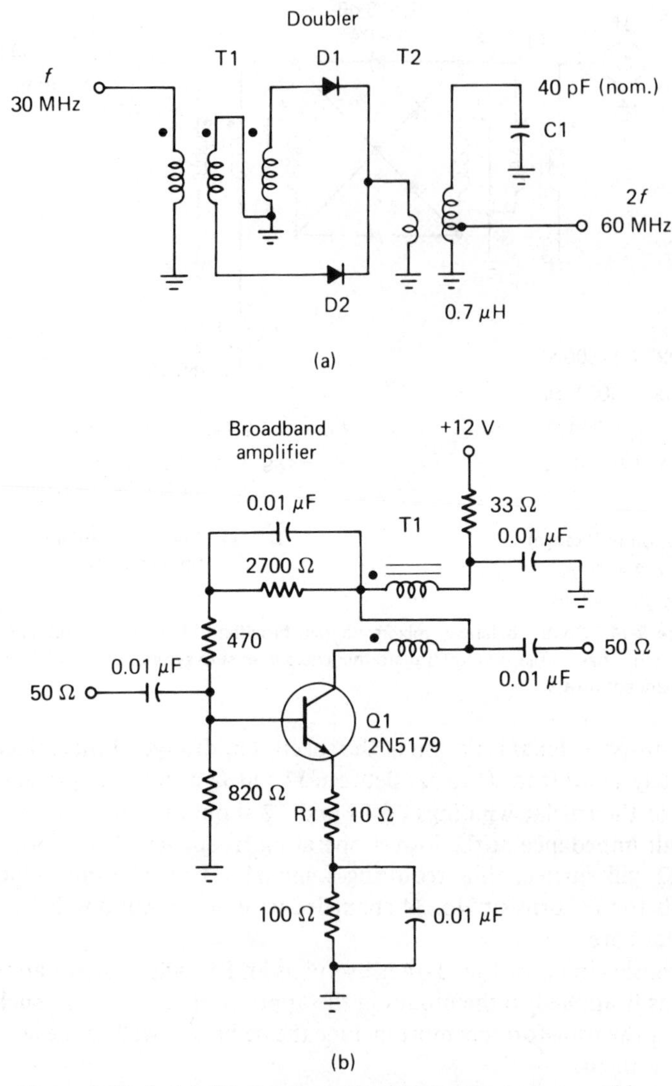

Figure 3-15 In (a), T1 and T2 are toroidal transformers as used in a balanced diode doubler. The broadband amplifier in (b) contains a 4:1 toroidal transformer (T1). The amplifier is suitable for use after the doubler in (a).

An iron core with a μ_i of 8 is a good choice at T2. If the core diameter is 0.5 in. (Amidon T50-6), a winding of 13 turns is required. No. 20 enameled wire is recommended. The output tap on the transformer secondary is chosen to provide an impedance match to the doubler load. If it is followed by a broadband amplifier like the one shown in Fig. 3-15b, the tap should be set for 50 Ω.

By using a broadband transformer in the manner illustrated in Fig. 3-15b, the amplifier will have a flat response over several octaves. One advantage of this circuit is that it has a 50-Ω characteristic at the input and output ports. It is a stable and predictable circuit by virtue of the shunt feedback from collector to base, and because of the degenerative feedback provided by R1.

Q1 is chosen for a high f_T if considerable amplifier bandwidth is required. The device specified has an f_T in excess of 1000 MHz. T1 is a 4:1 unbalanced-to-unbalanced transformer which steps the 200-Ω collector impedance down to 50 Ω. A suitable transformer would consist of 15 bifilar turns of No. 24 enameled wire on a 0.5-in.-diameter ferrite core which has a μ_i of 950. This would enable the amplifier to deliver a reasonably flat output from 3 to 30 MHz. A 125μ ferrite core would be suitable in this circuit if the transformer winding had sufficient turns to exhibit an X_L of 800 Ω at the lowest desired operating frequency.

3.2 Broadband Transformers

The broadband toroidal transformer is one of the most useful and commonplace components found in the RF communications industry. The form taken by these transformers is pretty much the designer's choice, but he or she may prefer the *transmission-line transformer* to the *conventional* type. Somewhat better efficiencies are available from the transmission-line versions, according to some designers. However, the impedance-transformation ratios obtainable with transmission-line transformers are restricted to specific values. The designer has relative freedom when working with conventional broadband transformers: He or she can develop any required turns ratio because the transformer windings are separate from one another. Transmission-line transformers, conversely, consist of bifilar, trifilar, or quadrifilar windings of some definite impedance. It is quite common to find these transformers with 25-Ω windings. The wires can be laid on the core in parallel, twisted a given number of times per inch, or they may consist of miniature 25-Ω coaxial cable.

Most broadband transformers are wound on ferrite cores because for a specified power-handling application the ferrites provide much higher permeabilities. This permits the use of considerably fewer turns of wire for ensuring reduced IR losses and adequate performance at the low end of the transformer bandwidth curve. In this instance we are considering the basic rule given earlier, where the transformer windings should have an X_L that is equal to or greater than four times the impedance to which the windings interface.

High-μ cores are ideal for broadband transformer use because the windings yield considerable inductance at the low end of the operating range, while the core material tends to "disappear" as the operating frequency is increased. Most transformers that are used from 2 to 30 MHz have a μ_i of 950. Higher values of permeability can be used, and on some occasions it is practical to employ cores with a lower μ_i (125). The latter, however, is applied more commonly at VHF, as are cores of relatively low μ.

Proper identification of the free ends of the windings on multifilar transformers can become an exercise in frustration. Therefore, it is prudent to use enameled wires of different color. Some wire manufacturers produce magnet wire that has red, green, or brown insulation. But, finding a small-quantity supplier is often impossible. Moreover, the minimum-billing fees imposed by some suppliers make it impractical to purchase small quantities of wire. An acceptable alternative to purchasing color-coded enamel wire is to use aerosol spray-can paint of various colors to set the windings apart visually. The grease-free lengths of wire are stretched out until taut. The spray paint is then applied and allowed to dry. Enamel paint is better than lacquer types of spray paint, as the coating will be less prone to cracking and falling off when enamel is used. Blue, green, red, and white are good colors to use for easy recognition.

3.2.1 Bifilar-Wound Transformers

Figure 3-16 shows four popular forms taken by toroidal transformers that employ bifilar windings. Since these are broadband transformers, the core material must be chosen to provide sufficient inductance for the lowest operating frequency. The core cross-sectional area must be appropriate for the power level of the circuit. For use from 1.5 to 40 MHz, cores with 125μ or 950μ are common.

A simple phase-reversal transformer is seen in Fig. 3-16a. This is a 1:1 impedance-ratio component. Both ports are single-ended. The windings can be formed by twisting the two conductors about 10 to 15 times per inch, then placing them on the core as one would do with a single conductor.

An unbalanced 4:1 transformer is illustrated schematically in Fig. 3-16b. This configuration is especially useful when stepping down from 50 Ω to the base of a solid-state, single-ended RF amplifier. Two of these transformers can be used in cascade, as shown in Fig. 3-17d, to obtain a 16:1 transformation. The 4:1 transformer is also useful when stepping up from a low collector impedance to 50 Ω or some other relatively low impedance.

A 4:1 balun (balanced to unbalanced) is presented in Fig. 3-16c. A popular application for this device is between a 50-Ω transmission line and the balanced 200-Ω terminals of an antenna. Similarly, it can be used between a 75-Ω coaxial line and the 300-Ω terminals of a folded dipole antenna. A number of TV receivers have used a solenoidal-wound rod type of balun for use between a 75-Ω coaxial feeder and the 300-Ω input terminals of the TV-set tuner. Toroidal-wound equivalents would be ideal for that use.

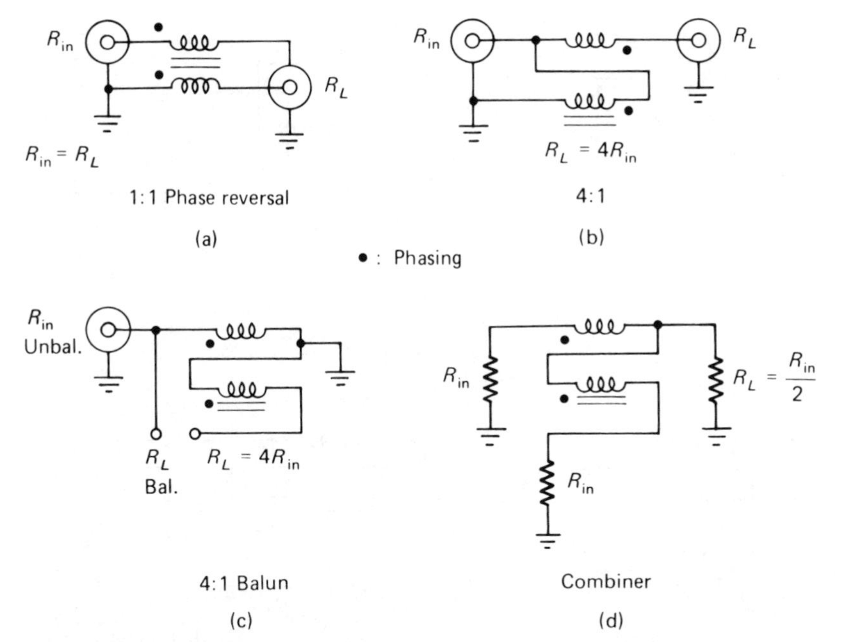

Figure 3-16 Four classic examples of broadband toroidal transformers. Each contains a bifilar winding.

A combiner transformer is illustrated in Fig. 3-16d. Two identical impedances can be combined (R'_{in} and R_{in}), as would be the case with two identical single-ended amplifiers, to provide a single output, R_L. For a matched condition, R_L would be equivalent to $R_{in}/2$. Thus, if two 50-Ω driving sources were used, the load at the remaining port would be 25 Ω.

3.2.2 Quadrifilar Transformers

A 9:1 unbalanced-to-unbalanced transformer is seen at in Fig. 3-17a. This configuration is often employed as a step-down or step-up coupler at the input or output of a single-ended solid-state amplifier. It is useful also in matching coaxial transmission lines to low-impedance antennas, such as a series-fed quarter-wavelength vertical antenna.

In situations where a 9:1 transformation is required for balanced-to-balanced terminations, the transformer of Fig. 3-17b is suitable. It is often used between push-pull drivers and push-pull amplifiers.

When the 1:1 transformer of Fig. 3-16a cannot be used, the balanced-to-unbalanced version of Fig. 3-17c will be suitable. This circuit is ideal for matching 75-Ω coaxial feed line to a 75-Ω dipole antenna.

The cascaded 4:1 transformers in Fig. 3-17d provide a 16:1 transformation ratio. T1 and T2 are identical, and are the same as the example given in Fig. 3-16b.

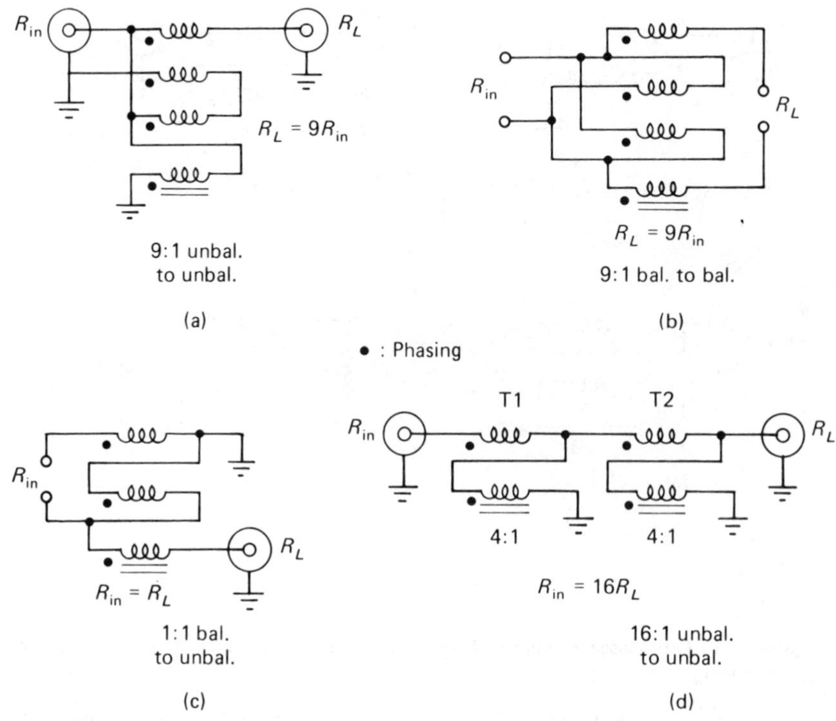

Figure 3-17 Multifilar windings are used on toroid cores to effect the broadband transformers in (a) through (c). Two 4:1 transformers can be used in cascade (c) to obtain a 16:1 impedance step-down.

3.2.3 Broadband Conventional Transformers

Although the consensus among designers seems to be that transmission-lines transformers are more efficient than "conventional" transformers are, the fixed-impedance ratios of the former are somewhat restrictive. We can avoid those sometimes confining parameters by utilizing the conventional type of transformers seen in Fig. 3-18. The physical format for a commonly used conventional broadband transformer is shown photographically in Fig. 3-19. A pictorial representation is provided in Fig. 3-20.

The simple illustration of Fig. 3-18 shows how two broadband conventional transformers might be used with a single-ended RF power amplifier. The 4:1 transformer of Fig. 3-16b could be used in place of T1 and T2 in this circuit, but only because the transformation is 4:1 at both amplifier ports. T1 and T2 of Fig. 3-18 would consist of single toroid cores made of ferrite. Since the operating frequency is fairly high, and because the large winding of each transformer should have an X_L of 200 Ω (1.06 μH at 30 MHz), a low-permeability material will suffice. A 0.5-in.-diameter core with a μ of 125 would be a suitable choice for T1. A pair of 1-in.-diameter

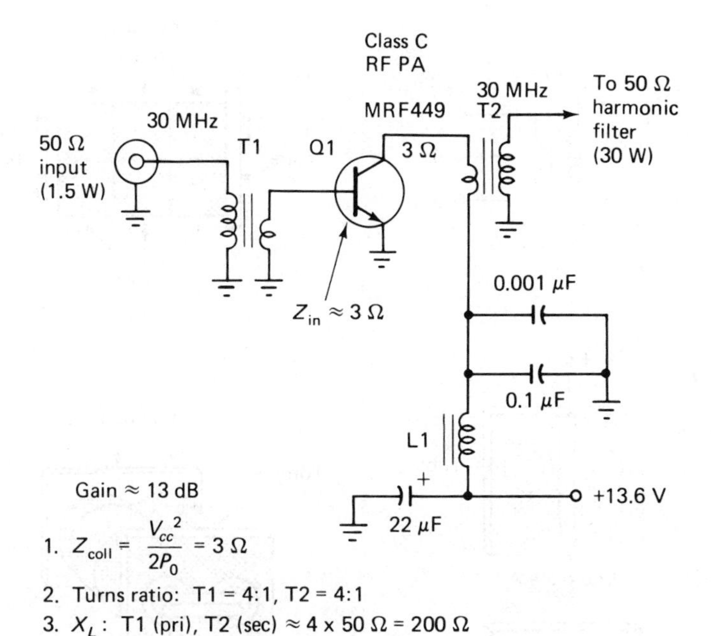

Figure 3-18 Conventional broadband toroidal transformers (T1 and T2) are shown in a Class C RF power amplifier.

Figure 3-19 Conventional transformer that might be found in a variety of RF power amplifiers.

ferrite cores ($125\mu_i$) can be stacked atop one another at T2 to provide ample B_{op} characteristics. L1 is a decoupling choke. A ferrite core capable of handling the dc current flow and able to provide a few μH of inductance will satisfy the requirement.

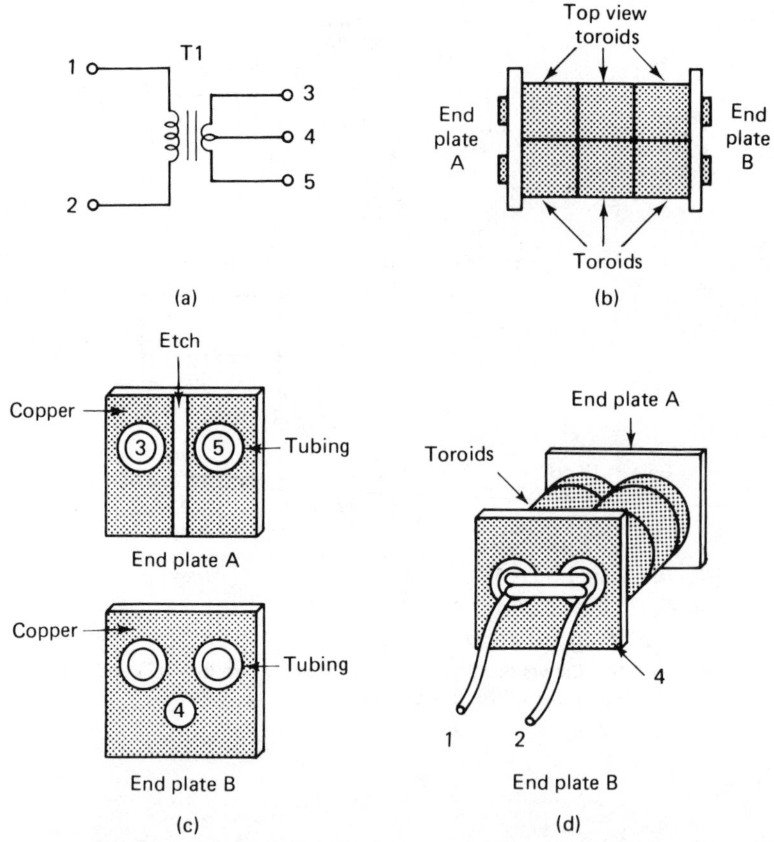

Figure 3-20 Circuit and structural details for a conventional transformer that uses rows of toroid cores. The end plates are made of printed-circuit-board material [(a) and (c)]. An assembled transformer is seen in (d).

When push-pull power amplifiers are used, as in the circuits of Figs. 3-21 and 3-22, the construction method seen in Fig. 3-20 is a good choice. Two rows of high-permeability ferrite cores ($950\mu_i$ typical for 2 to 30 MHz) are laid side by side as in Fig. 3-20b. The cross-sectional area of the stacked cores is selected for the power level of the amplifier. A piece of copper or brass tubing is passed through the center of each row of cores. The ends of the tubes are made common at end B of the transformer, but they are isolated from one another at end A, as seen in Fig. 3-20c. Printed-circuit-board material is suitable for making the end plates. End A is etched to prevent the tubing ends from being grounded or in contact electrically. This effective U-shaped tubing conductor constitutes a one-turn winding of the transformer. End plate B represents the center tap of that winding, or terminal 4 of Fig. 3-20a.

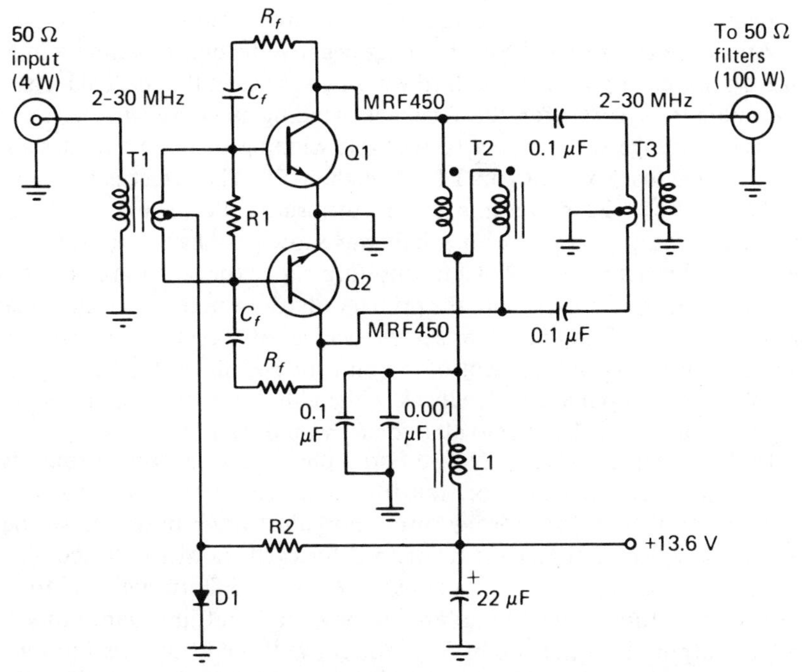

Figure 3-21 Circuit diagram of a typical push-pull solid-state RF power amplifier in which broadband toroidal transformers (T1, T2, and T3) are used.

The larger transformer winding is formed by looping the appropriate number of wire turns through the two pieces of tubing (terminals 1 and 2 of Fig. 3-20d). The turns ratio is set by considering the tubing as a one-turn winding, then using the proper number of wire turns through the core stacks. This conventional transformer is very compact and symmetrical.

Once the transformer is assembled it is a simple matter to affix it to the main etched-circuit board: end plates A and B are soldered to the appropriate circuit-board pads.

A circuit example of a broadband linear RF power amplifier is given in Fig. 3-21. T1 and T3 are conventional broadband transformers of the type presented in Fig. 3-20. T2 is a phase-reversal transformer that can be wound on a ferrite toroid. This transformer has a twofold advantage: operating voltage for the collectors of Q1 and Q2 is supplied through the windings of T2, but saturation is not a problem because the current flowing through the opposing windings produces opposite fields, thereby canceling. T2 also provides a 180° phase reversal, placing the collectors of Q1 and Q2 in balance.

Forward bias for the transistors is supplied through the secondary center tap of T1. D1 regulates the bias at approximately 0.7 V. R_f and C_f comprise the negative feedback networks for the amplifier. R1 swamps the secondary of T1 to aid stability at low frequencies. Decoupling choke L1 and the associated bypass capacitors also help to ensure stability.

An alternative method for obtaining negative feedback is shown in Fig. 3-22. In this example T1 has a third winding for use in the feedback circuit. The number of turns for the feedback winding is chosen to deliver the necessary voltage with respect to the base winding of T1. Some designers place the feedback winding in T3 rather than T1. The effect is the same.

T2 of Fig. 3-22 serves the same purpose as T2 of Fig. 3-21. This amplifier circuit is shown with a half-wave low-pass harmonic filter connected to the output of T3. Compensating capacitors (C_C) are connected across the secondary of T1 and the primary of T3—a necessary step in many broadband amplifiers that cover several octaves of frequency. R1 is included to permit precise adjustment of the bias for best linearity. R_e at Q1 and Q2 establishes degenerative feedback while aiding the dynamic balance of the two transistors. A typical value for R_e would be 1 Ω.

In the example of Fig. 3-23a we find a phase-reversal transformer used at T1 and a combiner type of transformer at T2. Both transformers are bifilar wound on ferrite toroid cores. For illustrative purposes we have specified a collector characteristic of 15 Ω for each transistor. T2 combines the two impedances to provide an output level of 7.5 Ω. In such an instance the harmonic filter would be tailored to have an input impedance of 7.5 Ω and an output characteristic of 50 Ω for conventional transmission lines.

The hybrid couplers shown in Fig. 3-23a and b (T1 through T3) can be wound toroidally or on balun types of magnetic cores. They are used in many designs to combine the outputs of individual power amplifiers, or to combine the outputs of groups of power modules. It has become standard practice to use this style of transformer at power levels as great as 1 kW throughout the high-frequency bands.

Hybrid transformers can be reversed and used to split a single power source into two or more branches of equal phase and amplitude. When they are constructed and used correctly, they provide isolation amounts of between 30 and 40 dB from 1.8 to 30 MHz between the two like power sources. This isolation enables the system to continue operating, even though one of the power sources may fail. Although failure of one amplifier will result in reduced power output, a constant load impedance will exist. This will protect power transistors from damage and will preserve the linearity of amplifiers which operate in that mode.

The balancing resistors of Fig. 3-23 (R1 and R2) keep the VSWR at a low value even if one of the amplifiers fail. Each balancing resistor must be twice the ohmic value of the driving source. The resistor should be rated for at least 0.25 of the output power of the total system, and it must be the noninductive type. Granberg treats this general subject in depth in Motorola *AN-749* [4].

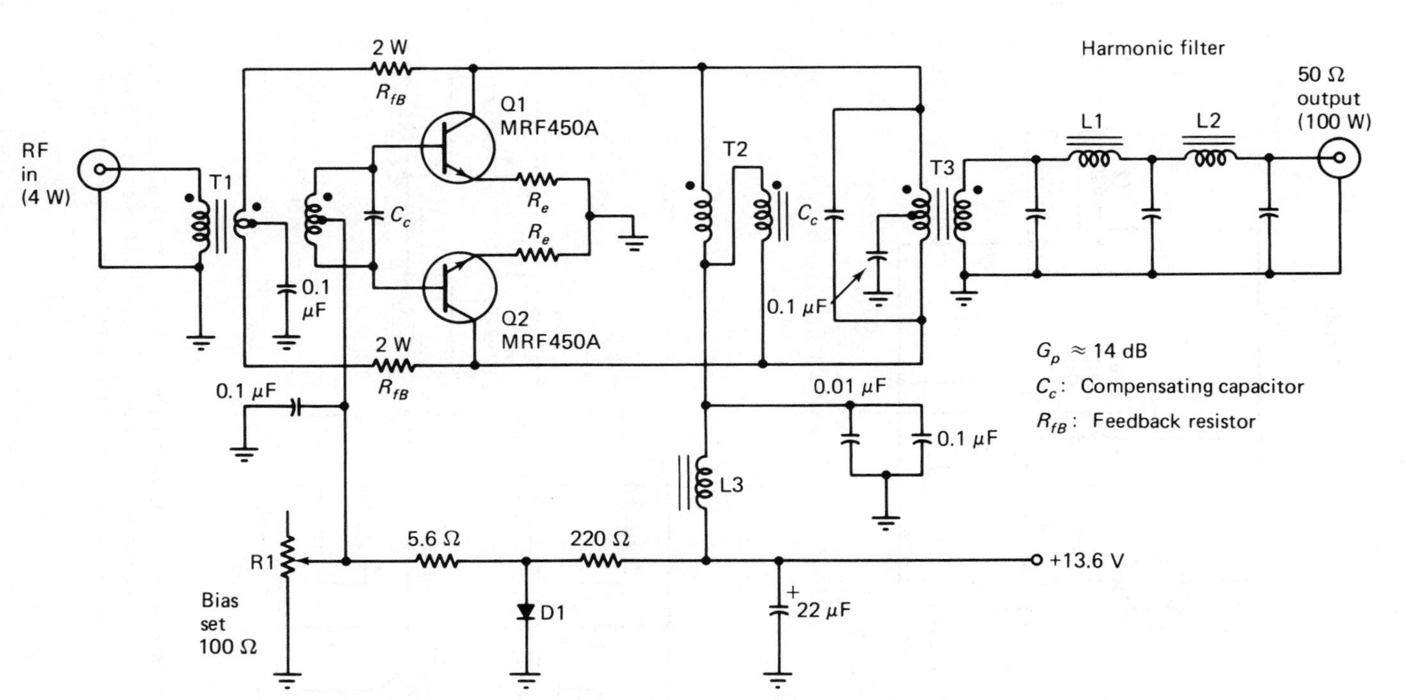

Figure 3-22 T1 has a third winding which is used in the negative-feedback loop of this amplifier.

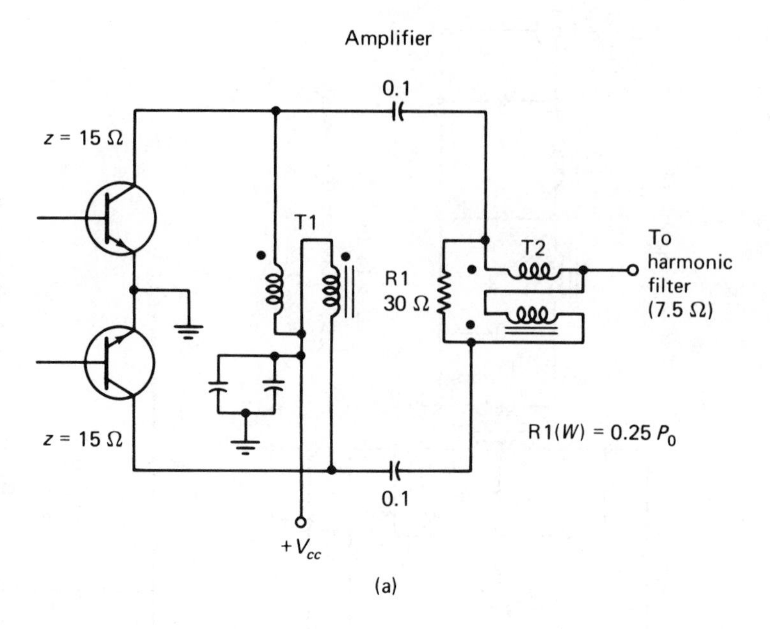

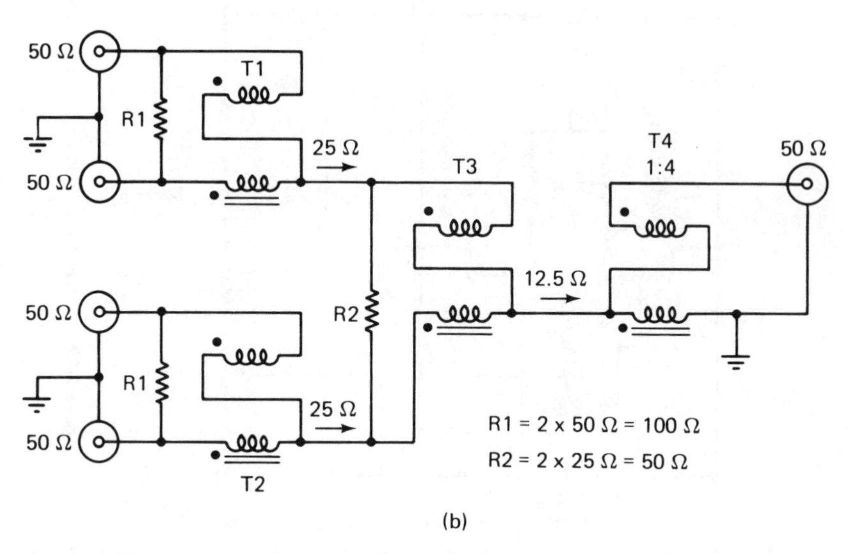

Figure 3-23 Circuit that shows toroidal transformer T1 as a phase-reversal collector choke. T2 in this example is employed as a combiner transformer. In (b) is seen a complex hybrid combiner which consists of four broadband toroidal transformers (see the text).

3.2.4 Variable Impedance Matching

The conventional-transformer concept we have been discussing can be applied as shown in Fig. 3-24a to effect a matched condition between, say, 50 Ω and an unknown load, R_L. Sufficient reactance must be present in the windings to ensure minimum losses at the lowest operating frequency of the transformer. The secondary winding of the toroidal broadband transformer can contain as many tap points as the designer deems necessary. T1 can be inserted in the line between a known R_{in}, inclusive of an SWR indicator, then the taps on the secondary winding are varied until the lowest SWR is noted—preferably an SWR of 1. This type of transformer is especially useful when matching a 50-Ω transmission line to a quarter-wavelength vertical antenna: the feed-point impedance is usually unknown, owing to the quality of the ground (conductivity) in the vicinity of the antenna. Values from a few ohms up to 30 Ω are typical. The core material is usually ferrite (μ_i of 125 for HF-band applications) and the cross-sectional area is chosen for the power level in use, in accordance with the B_{max} rules. Teflon-insulated wire is recommended for the transformer windings.

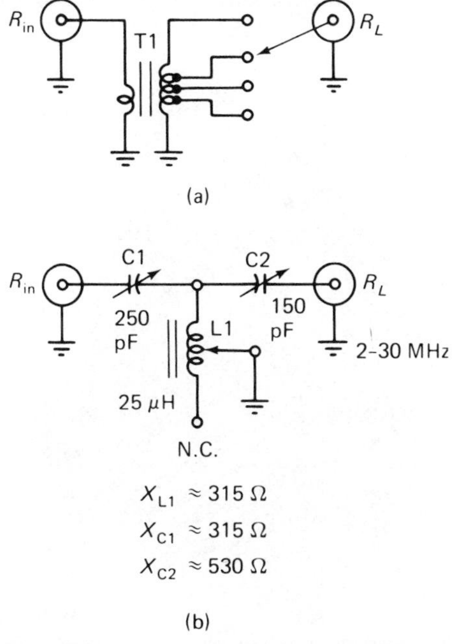

Figure 3-24 The transformer in (a) has a tapped secondary to permit matching various impedances to a specified primary load. In (b) is a T-matching network that contains a narrow-band format. L1 can be built as a nonresistive rheostat, using a toroid core as the foundation (see Fig. 3-25).

A narrow-band type of matching circuit is illustrated schematically in Fig. 3-24b, and photographically in Fig. 3-25. L1 is a toroidal inductor which has been wound on a large powdered-iron core. The completed assembly resembles a power rheostat in appearance. Bare wire is used for the winding, and care is taken to prevent the turns from shorting together. The movable arm of the variable-inductor assembly selects one turn at a time. It must not short any two turns when it is in position. The lower end of L1 (marked "N.C.") is left open to prevent a shorted-turn condition.

Under certain conditions the circuit is a high-pass T network. Depending upon the values of impedance being matched, a bandpass response can result. C1, C2, and L1 are adjusted in sequence to obtain the lowest reflected power possible. In terms of network loss, C2 should be set for the maximum amount of capacitance that will permit a matched condition with respect to the settings of C1 and L1. Since considerable RF voltage can be present when R_L is of high impedance, insulating material of high dielectric quality should be used between the winding and the toroid core. Heavy-gauge silver-plated wire is ideal for the winding of L1. Rule-of-thumb reactances are listed for C1, C2, and L1 of Fig. 3-24b.

The limitation of the inductor shown in Fig. 3-24b is that the resolution

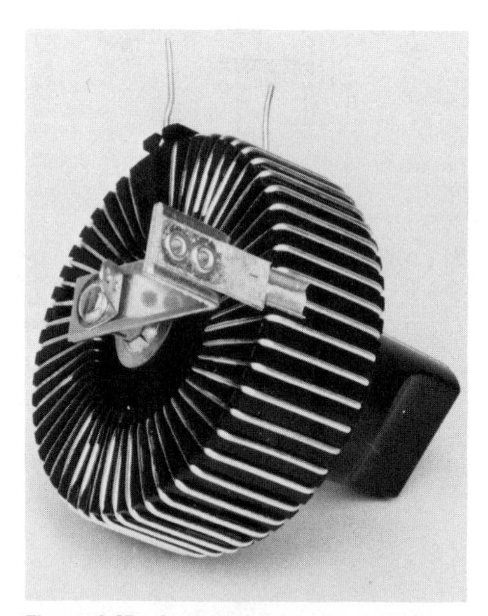

Figure 3-25 Commercially made toroidal type of variable inductor. This unit is manufactured by Ten Tec Corp. for use in their antenna Transmatch. The wiper is built so that it contacts one wire turn at a time, thereby preventing the shorted-turn effect.

of inductance must be in one-turn increments. Under some circumstances this can present a problem, especially if the core has high permeability: one turn can represent considerable inductance. A continuously variable inductor can be realized by adopting the technique seen in Fig. 3-26. One half of a powdered-iron toroid core is obtained by sawing the toroid in half. The core half is affixed to an insulated arm which is connected to a ¼-in.-diameter brass shaft, as shown. A semicircular air-wound coil is mounted opposite the core half. It has an ID that permits free passage of the core half when the tuning shaft is rotated. The overall assembly is similar to that of a small potentiometer such as one might use as an audio-gain control. For wide-range inductances the designer can switch additional inductance in series with the variable one, or fixed-value inductance can be placed in parallel with L1 to secure lower minimum-inductance values. This type of unit is ideal for remote tuning by means of a motor. It is useful also as a panel-mounted control. Circuit connections are made to points X and Y.

A versatile variable-impedance broadband matching transformer is presented in Fig. 3-27. It was developed by J. Sevick of Bell Labs. It consists of a fairly large toroid core on which a quadrifilar winding is placed. Impedance transformations of 4:1, 9:1, and 16:1 are achieved easily with the basic configuration shown. Additionally, points X, Y, and Z can be attached to the windings at the appropriate locations to obtain ratios as low as 1.5:1. When matching between relatively low impedances, say 50 Ω to values

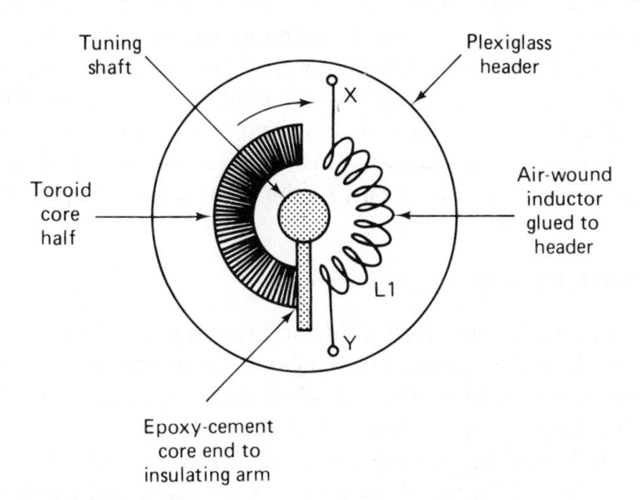

Figure 3-26 How a variable inductor can be made from one-half of a toroid core. This scheme was devised by the author in 1970 for use in a low-power antenna Transmatch for field use. As the core half enters the 180° air-wound coil, the inductance increases. Mechanical details were developed by A. Pfieffer of Old Lyme, Connecticut.

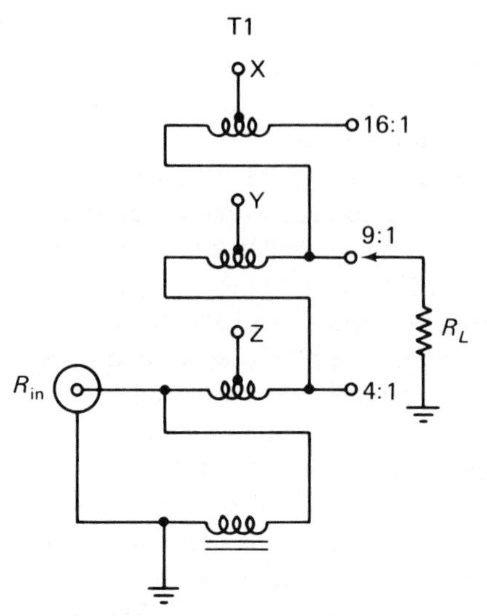

Figure 3-27 Circuit diagram of a wide-range toroidal matching transformer which will provide ratios from 1.5:1 up to 16:1.

below 50 Ω, the transformer can handle up to 1 kW of power. This is based on a toroid-core diameter of 2.5 in. and a winding that employs No. 14 or larger enameled wire. For the medium and high frequencies, core permeabilities of 125 and 40 are common. Cores with 950 μ_i would be suitable for the broadcast band and lower. A 10-turn quadrifilar winding on a 2.5-in.-diameter toroid core ($\mu_i = 125$) would be satisfactory for use from 2 to 30 MHz. The transformer leads should be kept as short as possible to minimize unwanted inductances. This is especially significant at very low impedance levels at the upper part of the high-frequency spectrum.

3.3 Antenna Loading

An interesting application for toroids or ferromagnetic sleeves is illustrated in Fig. 3-28. A dipole antenna can be made significantly shorter while maintaining the proper half-wavelength electrical characteristic by loading it with ferrite. Experiments conducted in the VHF spectrum indicated that physical-length reductions as great as 2 were entirely possible by using low-μ ferrites over the two halves of the dipole. The voltage and current distribution of the system remained linear, and no measurable degradation in efficiency was observed. The larger the cross-sectional area of the toroids, the greater the size reduction. A similar result would be had by increasing the μ_e of the ferrite material, consistent with suitable Q versus frequency.

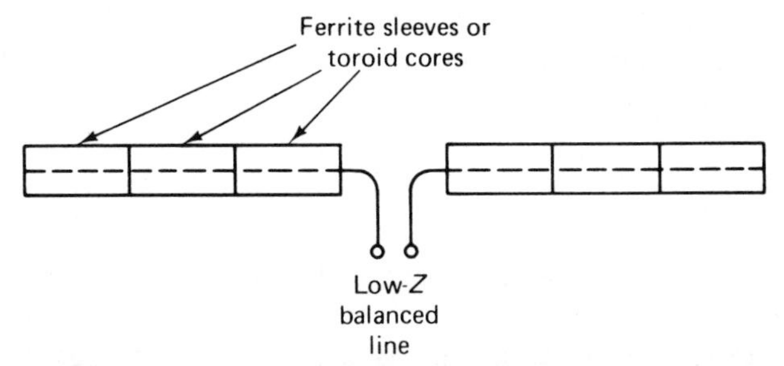

Figure 3-28 Ferrite cores or sleeves can be placed over the conductors of a dipole antenna to decrease the antenna length. Field-strength measurements indicated no measureable degradation in antenna performance as a result of the loaded dimensions.

3.4 Toroidal Distributed Capacitance

When designing the transformers discussed earlier in this chapter, it is important to consider the distributed capacitance of the inductor or transformer. Many things relate to the distributed capacitance. Part of the problem results from the differential voltage between adjacent turns. The amount of capacitance is dependent upon the type of insulation (dielectric) on the wire and the core material. If moisture is present—even in small quantities—it also affects the capacitance. This parasitic capacitance can be regarded as a fixed-value capacitor in parallel with the toroid winding. The winding inductance and stray capacitance establish the self-resonant frequency of the inductor.

If the operating frequency of the toroid is too near that of self-resonance, the circuit Q is lowered and the apparent inductance increases. This is particularly prominent when high operating frequencies are used and when the toroid has high inductance. To illustrate the effects of this unwanted condition, assume that the operating frequency is 10% of the self-resonant frequency. The resultant actual Q will be roughly 99% of the computed value. The condition worsens as the operating frequency is brought closer to the self-resonant one: a differential of 50% will degrade the calculated Q to 75%.

Reduction of the capacitance can be realized by adopting specific winding techniques. For example, a scramble-wound core that occupies 360° of the toroid will have relatively high stray capacitance. A core with, say, 900 turns will exhibit a parasitic capacitance of 40 to 70 pF. This can be improved by using fewer degrees of core circumference, such as 325°.

Some designers use a progressive or modified-bank style of winding. A gap of ⅛ to ¼ in. is maintained between the ends of the winding. The stray capacitance will be reduced some 50% over the scramble-wound 360° type of toroid winding. It should be noted that the encapsulation material used for potting toroidal inductors and transformers will add somewhat to the distributed capacitance.

3.5 Power Supplies

Toroids, as well as U cores, pot cores, and E cores, are used extensively in transformers for dc-to-dc converters and dc-to-ac inverters. The principle of operation for these power supplies is relatively simple: The transformer is wound on a core that will provide a hysteresis loop whose shape is nearly that of a rectangle, as is seen in Fig. 3-29. This type of transformer is found in Fig. 3-30 (T1).

Q1 and Q2 of the same circuit operate as switches. When one is off the other is on, and vice versa. So that we may illustrate the operational concept, let us assume that Q1 is conducting and that Q2 is in cutoff. This condition places the input voltage across the top half of N_{pri} of Fig. 3.30. As this occurs, a voltage is induced in all the T1 windings. When Q1 conducts, instantaneous current and voltage peaks reach maximum in the transformer windings. This condition will be maintained until the core of T1 saturates. At this time the flux change rate will drop to zero, and the induced voltages will decline to zero. As this happens, the base drive disappears from Q1. At this period the current decreases, causing the flux to increase in the opposite direction. Since a voltage of opposite polarity is induced in the windings of

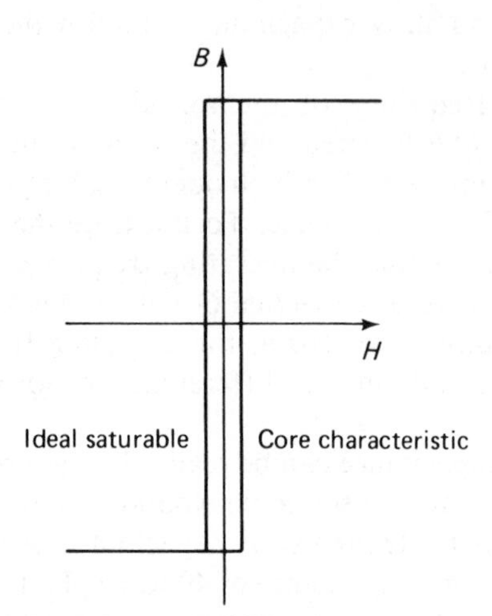

Figure 3-29 An ideal hysteresis loop for a saturable core would be a perfect rectangle, as shown here. (Courtesy of Magnetics Division of Spang Industries, Inc.)

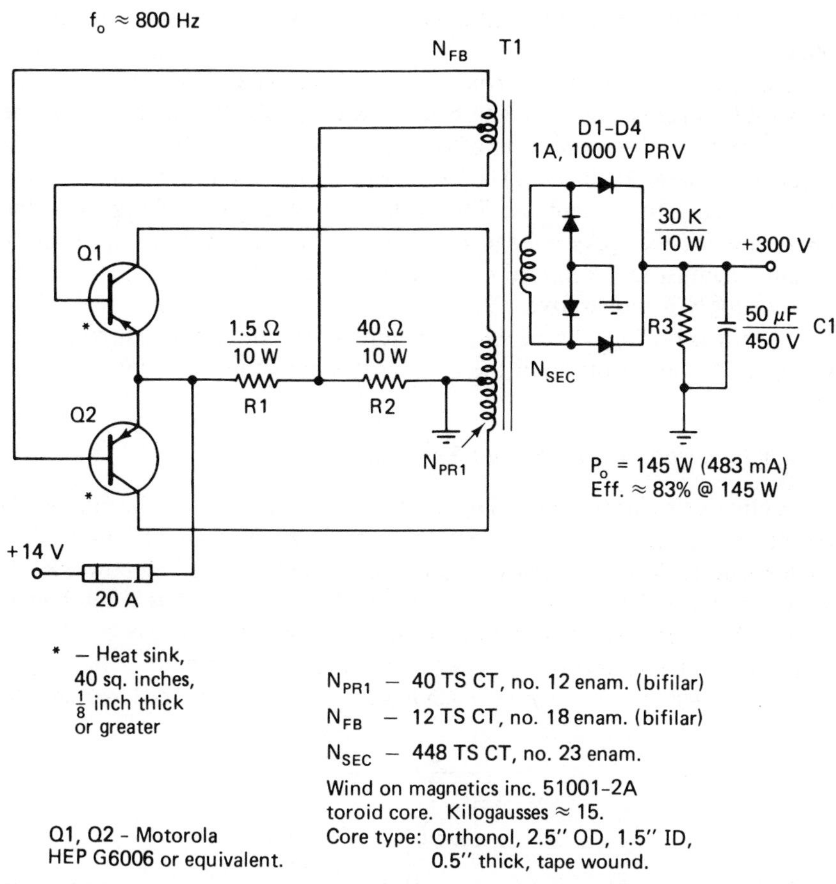

Figure 3-30 Basic circuit for a dc-to-dc converter in which a saturable transformer (T1) is used. This circuit is patterned after one that was published by Delco Division of GM Corp. in *AN-1B,* an application note.

T1, Q2 goes into conduction. From this point onward, the cycle keeps repeating.

A square wave will appear across N_{sec} of Fig. 3-30. The frequency of the Q1, Q2 switching action is determined by number of primary turns on T1, the dc operating voltage, and the flux characteristics of the toroid core. Hence,

$$N_{pri} = \frac{E_{bb} \times 10^8}{4B_s A f}$$

where E_{bb} is the supply voltage, N_{pri} the number of turns in *one-half* the primary winding, B_s the saturation flux density in lines per square inch, A the core area in square inches, and f the frequency of switching in hertz. The frequency, from no load to full load, should not change more than 5% if the design is done correctly.

Oscillation is started by virtue of R1 and R2. These resistances provide forward bias for Q1 and Q2. This causes the transistors to operate above the low-current, nonlinear portion of the curve. A full-wave bridge rectifier is formed by using D1 through D4, as shown. R3 is a minimum-load resistor. It is necessary to prevent the 50-μF filter capacitor from charging to the peak value of the spike voltage which is present. It serves also as a safety measure against shock hazard by bleeding the dc voltage from C1 when the power supply is inoperative.

The turns on N_{sec} can be varied in number to obtain the desired output voltage. The power rating of the composite supply must be kept in mind if this is done.

3.5.1 Using Power FETs as Switches

A circuit similar to that of Fig. 3-30 is given in Fig. 3-31. The major difference is that Q1 and Q2 are VMOS power FETs. The use of power FETs enables the designer to employ smaller components for a given power level. This is because the upper frequency limit of FET switches is much higher (100 MHz or greater), thereby permitting switching frequencies of 50 kHz or greater. Smaller toroid cores can be used and the oscillator efficiency is high. The filter capacitor can be much lower in value than at the lower switch-

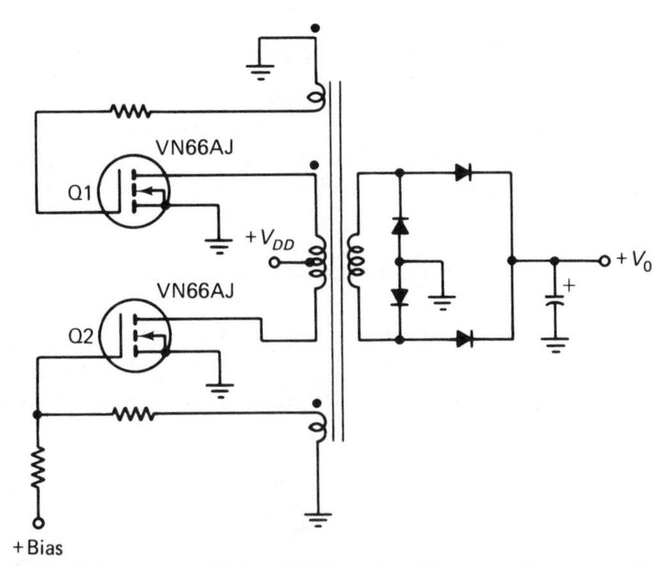

Figure 3-31 Power FETs (VMOS) are ideal for use with a saturable transformer. The basic circuit seen here is capable of highly efficient operation.

ing frequencies of a few hundred hertz to 3 kHz. The Siliconix VN66AJs shown are but one type of VMOS power FET suitable for high-current, high-speed switching.

Other advantages are offered by enhancement-mode power FETs: they are not subject to destruction from thermal runaway or secondary breakdown. They cannot be destroyed when operated into an open or shorted load. Of more than casual interest to the designer is the low input and output capacitance characteristics of these FETs: the input amount is typically about 50 pF and the output is on the order of 35 pF. This explains the fast switching time that they will provide. Power FETs are somewhat more efficient at the higher operating voltages, so the circuit shown in Fig. 3-31 would be ideal for use from a 24- to 28-V dc supply. Dots indicate the polarity of the feedback and primary windings of T1. The same type of core material specified for T1 of Fig. 3-30 (Magnetics, Inc., Orthonol) is recommended for the lower oscillator frequencies. A ferrite toroid core would be more appropriate for high operating frequencies (10 to 50 kHz).

Figure 3-32 shows how an Orthonol core compares to others which are used for dc-to-dc converters and dc-to-ac inverters. The *B–H* loop closely approaches the classic rectangle discussed earlier.

The core loss versus flux density and frequency is shown in the curves of Fig. 3-33. The curves are for 2-mil Orthonol such as that used in T1 of Fig. 3-30.

3.6 Core Doping

There is seldom an application for toroidal inductors or transformers in which some form of protection against dirt, abrasion, and moisture is not used. The final component is usually coated with a low-loss, durable protective lacquer or compound. This can take the form of simple impregnation of the core and winding with coil dope, or in a more elaborate situation the assembly may be encapsulated by means of a potting compound.

There are a number of reasons why we might want to seal the winding of a toroidal-wound component. First, movement of the coil or transformer turns has a significant effect on the effective inductance. This is especially true when a single-layer winding is applied to a core and when there is space between the turns. Any change in inductance can have a serious effect on circuits which perform as oscillators or filters. In the case of an oscillator, the frequency stability can be substandard if the turns are allowed to move. Furthermore, if the overall piece of equipment in which the toroid is used happens to be in an environment where vibration takes place, frequency modulation may result when the turns of the coil vibrate. A narrowband filter will not hold alignment if the inductors are incapable of being stable.

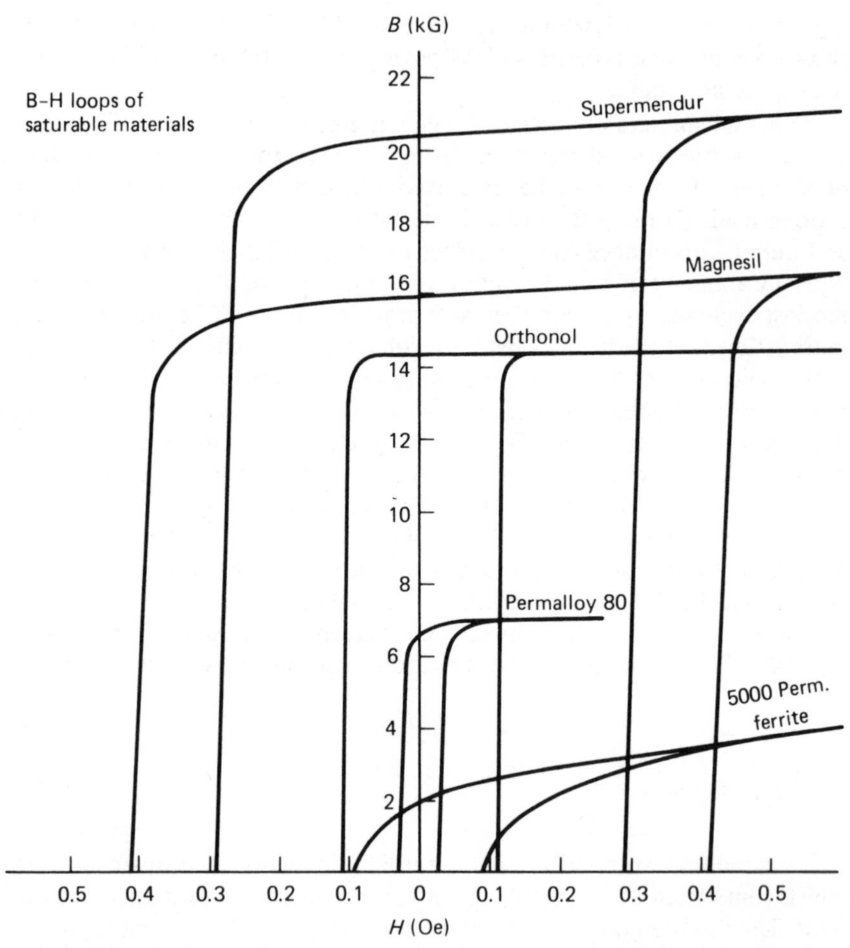

Figure 3-32 Comparison of core types for use in saturable reactors. The Orthonol core has an almost ideal rectangular hysteresis characteristic. (Courtesy of Magnetics Division of Spang Industries, Inc.).

Therefore, it is prudent to affix the coil turns to the core material in filters as well as oscillators.

3.6.1 Simple Doping Methods

Figure 3-34 shows two broadband, trifilar-wound toroidal transformers being used in a doubly balanced passive mixer. The transformer windings have been sealed by means of polystyrene-base coil cement (Polystyrene Q-Dope). This is a high-dielectric substance which has a tough consistency. Three coats of cement were applied to the toroids shown in the photograph. The last coating was allowed to flow through the core center and around the

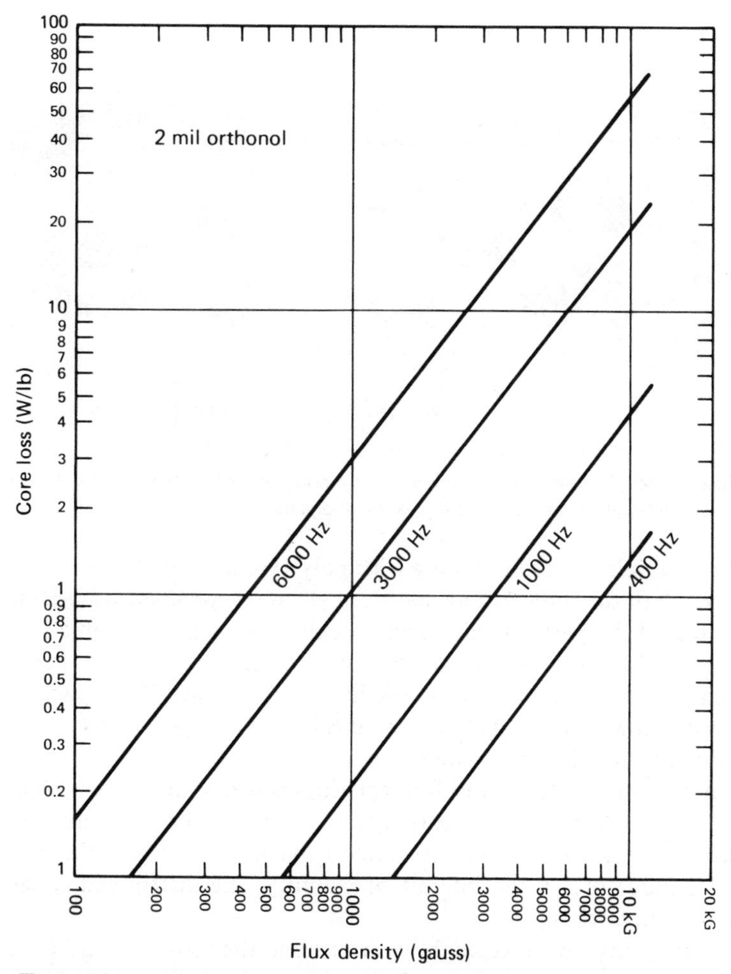

Figure 3-33 Flux density and frequency of operation versus core loss is illustrated here. These curves are based on 2-mil Orthonol core material. (Courtesy of Magnetics Division of Spang Industries, Inc.).

outer edges. This helps keep the transformers in place on the printed-circuit board. There is no copper conductor on the component side of the printed-circuit board. Had there been, each transformer would have been elevated at least ¼ in. above the copper ground plane, using an insulating pedestal. Such an elevating insulator serves two purposes: it prevents abrasion of the windings and consequent short-circuit conditions. Also, placing the transformer somewhat above the copper ground plane reduces unwanted capacitive effects between the windings and ground. In a symmetrical circuit, such as a balanced mixer, all stray capacitance and inductance must be kept to a minimum.

Figure 3-34 The two toroids shown in this module are trifilar wound. They have been doped with polystyrene Q-dope to seal them against moisture.

An example of toroid doping with polystyrene cement is offered in Fig. 3-35. Each of the three toroids has been given several coats of Q-Dope. The large toroid is part of the tank circuit of a FET variable-frequency oscillator. Directly below it are three temperature-stable polystyrene capacitors. They are part of the oscillator *LC* circuit. The capacitors have also been affixed to the pc board by means of Q-Dope, thereby preventing unwanted instability from vibration.

Another desirable end result of coil doping is that moisture is kept out of the windings. As we learned earlier in the chapter, the moisture content in the air has an effect on the distributed capacitance between the turns of a toroidal inductor. In the interest of stability, the distributed capacitance must remain unchanged.

In situations where we might wish to stand the toroid on end for ease of mounting and to lessen unwanted capacitive effects, the method illustrated in Fig. 3-36 is suggested. The toroid is first "gunked" with Q-dope and allowed to dry. Then a generous dab of GE Silastic compound or RTV sealant is dropped on the printed-circuit board. The toroid, which has already been soldered to the circuit board, is allowed to "float" in the cement as shown. Once the compound has set, the toroid is held securely in place. The compound also functions as an effective shock absorber.

3.6.2 Potting Technique

It is standard practice to imbed low-frequency toroids in potting compound. Many Mil-Spec products require potting procedures to protect the component from moisture, oil, dirt, and fungus. Although the potting compound always has some effect on the electrical characteristics of a toroidal

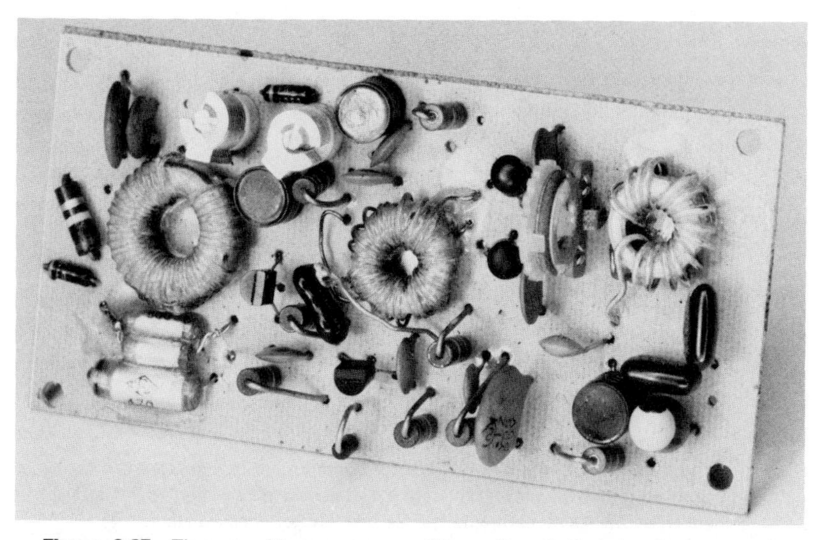

Figure 3-35 Three toroids are seen on this pc board. Each has been coated generously with Q-Dope to affix them to the circuit board and protect the windings from abrasion and moisture.

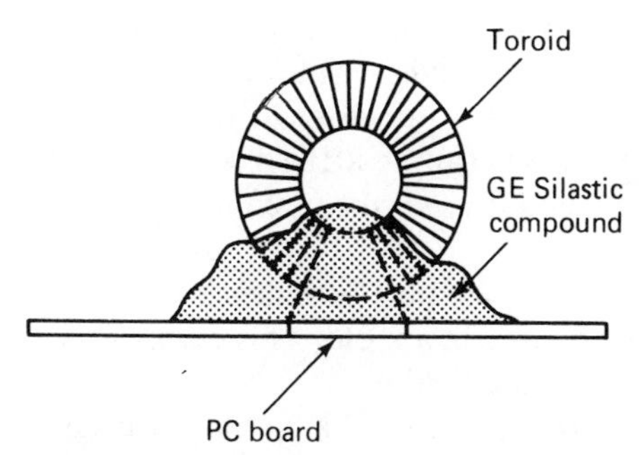

Figure 3-36 A toroid can be edge-mounted by floating it in a dab of silicone cement such as GE Silastic or RTV sealant. The cement has a rubber-like consistency when dry, thereby functioning as a shock mount.

inductor or transformer, the problem is of minor consequence at the lower frequencies.

Various kinds of chemical compounds are available, and each has a specific stability factor and dielectric quality. Once the appropriate substance is selected by the designer, the toroid can be encapsulated in the manner shown at Fig. 3-37. Although the illustration depicts but one of many potting formats, it is one of the most common ones used by the industry.

The inductor or transformer is wound, tested, then connected to a piece

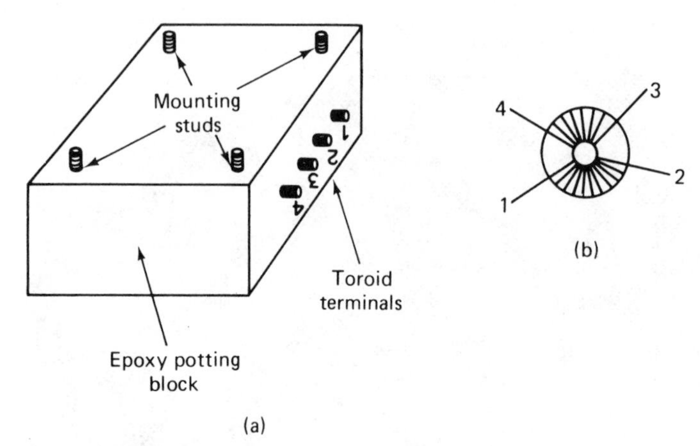

Figure 3-37 The longevity of toroidal transformers or inductors can be enhanced greatly by placing them in a potted assembly of the type shown here. Most Mil-Spec contracts call for this kind of encapsulation.

of circuit board that has been fitted with solder terminals. A mold of suitable dimensions is chosen next. It is coated internally with silicone grease to prevent the potting compound from adhering to it. The toroidal assembly and terminal board is inserted in the mold, then the potting compound is poured. Mounting studs are also contained on the printed-circuit-board header if mounting is to be done at the terminals. Alternatively, as seen in Fig. 3-37, the studs can be affixed to one side of the finished product. If the latter method is used, the mold must have holes on one wall to accommodate the studs.

Many potting compounds must be baked for a specified period of time after they have been poured in a mold. This curing process ensures a stable and reliable component for field use.

Various dyes are available for making the encapsulating material practically any color the designer specifies. Some casting resins which are sold in hobby stores are entirely suitable for one-shot laboratory experiments with encapsulation. Few of them require oven baking, and dyes can be mixed into the resin as desired.

REFERENCES

[1] **John D. Lenk,** *Handbook of Modern Solid-State Amplifiers.* Englewood Cliffs, N.J.: Prentice-Hall, Inc., 1974.

[2] **M.F. "Doug" DeMaw,** *ARRL Electronics Data Book.* Newington, CT: ARRL, Inc., 1976.

[3] **Wes Hayward and M.F. "Doug" DeMaw,** *Solid State Design for the Radio Amateur.* Newington, CT: ARRL, Inc., 1977.

4

BEADS, SLEEVES, AND POT CORES

Although we covered the subject of toroids in Chapter 3, the ferrite beads and sleeves treated in this section are similar to their doughnut-shaped cousins, the toroids. The principal difference between beads and toroids is that the former are of a different aspect ratio, physically, as can be seen in the photograph of Figure 4-1. The diameter/length ratio of beads is greater than for toroids. Similarly, the ID/OD ratio is greater, and beads

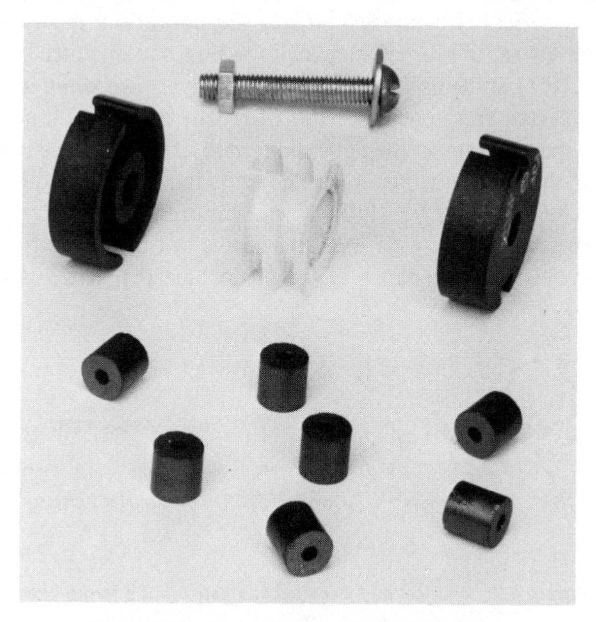

Figure 4-1 Pot core and some ferrite beads.

are generally much smaller overall than the typical toroid core. Finally, the applications to which beads are put are somewhat apart from those of toroids.

We can envision the ferrite sleeve as a long ferrite bead which has a relatively thin wall, comparatively speaking. It can be used for many of the tasks assigned to the bead, as we shall learn in this chapter. For most circuit applications the bead, sleeve, and pot core are used to surround the conductor (single wire or coil), whereas the conductor is on the outer surface of a toroid core. Thus, the bead, sleeve, and pot core not only increase the inductance of the conductor, but also serve as shield material for the inductor.

Pot cores and cup cores are two-piece units. One-half of the ferrite cylinder contains an insulating bobbin on which an inductor or transformer winding is placed. The remaining core half fits over the bobbin and mates with the first half of the core. The two sections are tightened together by means of a nonconductive screw-and-nut assembly. The resultant device is an enclosed, shielded transformer or inductor.

4.1 Properties of Beads

The physical and electrical characteristics of the ferrite bead are shown in Fig. 4-2. A blank bead is depicted in part (a). Part (b) shows a single-wire conductor passed through the bead. Part (c) indicates the existence of *R* and *L* components that result from the assembly seen at *B*. When a ferrite bead is inserted in series with an ac voltage, the condition shown in Fig. 4-3 prevails. Therefore, the impedance exhibited by one or more beads slipped over a conductor can be used to considerable advantage when it is desired to suppress a selected frequency or band of frequencies. The amount of suppression is dependent upon the μ_e of the ferrite, the size of the bead, and the frequency of the suppressed energy. The impedance presented by the ferrite material can be increased by winding two or more turns of wire through the bead. This is seen in Fig. 4-4a. An alternative but more costly method for increasing the series impedance is shown in part (b), where two or more

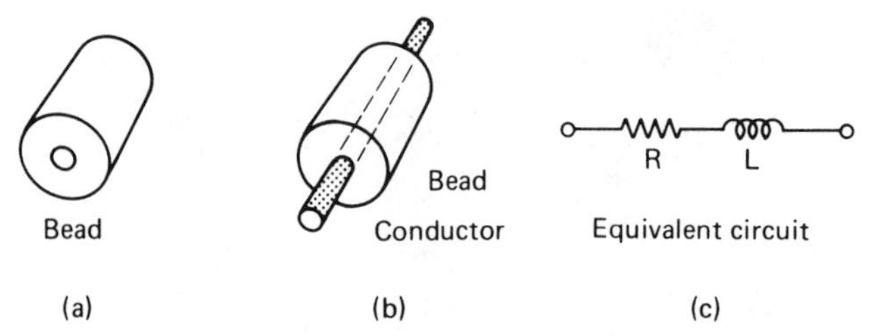

Figure 4-2 Physical and electrical comparison of a ferrite bead.

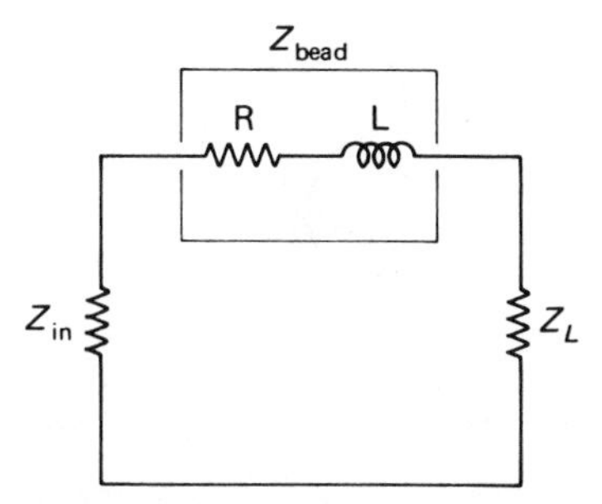

Figure 4-3 Schematic representation of a ferrite bead when inserted in a signal path.

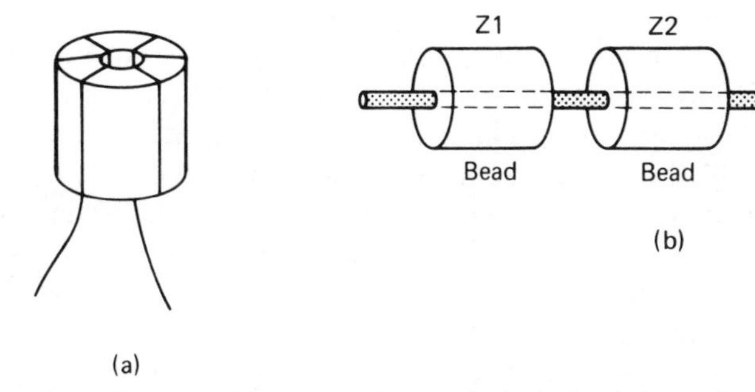

Figure 4-4 The effective inductance of a ferrite-loaded impedance can be increased (a) by looping one or more wire turns through a single bead, or (b) adding beads in series on a single conductor.

beads can be used in a string to exceed the inductance of a single component.

An increase in bead/conductor inductance can be achieved easily by utilizing a slightly different form of bead—one which has two or more holes through it. Figure 4-5 contains examples of three popular multihole beads. The device seen in part (c) is called a *balun* core. The two holes are of somewhat larger diameter than those found in ferrite beads. The bead shown in Fig. 4-2a, plus those in Fig. 4-5a and b, are not suitable for passing more than one turn of medium-gauge wire through their bores. The balun core in Fig. 4-5c is ideal when several turns of wire must be wound through the core holes.

Some manufacturers will supply ferrite beads that are assembled on No. 22 AWG tinned copper wire. Fair-Rite Products Corp. is one such fabricator. The beads can be supplied, taped, and reeled for use by automatic insertion apparatus for printed-circuit boards. They are supplied in accordance with EIA Standard RS-296-C.

Ferrite beads are available in a variety of permeabilities. Material with μ_i

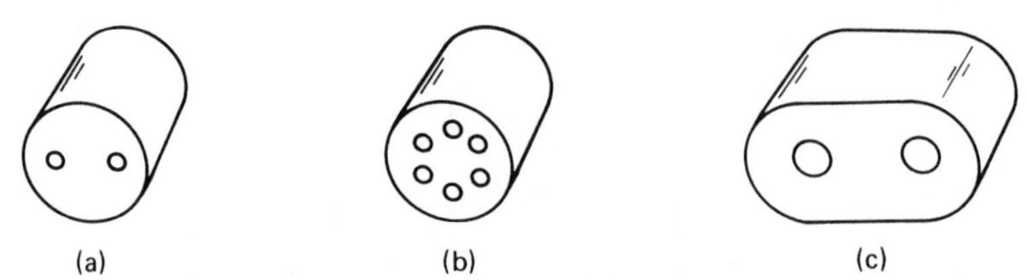

Figure 4-5 Ferrites are available for use as baluns or single impedances in the forms seen at (a) through (c). The number of holes will depend on the application.

values of 2500 and 5000 is excellent for frequencies up to 40 MHz. From 40 to 200 MHz a bead with 850 μ_i is recommended. Above 200 MHz it is best to employ bead material that has a 250 μ_i characteristic.

The absorption of ferrite beads can be enhanced measurably by using a bead and a capacitor in series, but with the load in parallel with the capacitor (Fig. 4-6). The resultant damping value is equivalent to the bead impedance multiplied by the reactance of the capacitor. It can be seen from this that the three beads shown in Fig. 4-6 would by themselves exhibit a series impedance of 75 Ω at 2.2 MHz. Through the addition of C1, the effective damping impedance has been elevated to 5422.5 Ω. The benefits of this method are especially significant when it is necessary to filter unwanted frequencies from power leads. Attenuation amounts of 50 dB or greater are commonplace at VHF and higher. in RFI and EMI filtering applications it is standard practice to use feedthrough capacitors in combination with ferrite beads at the entrance and exit points of shield enclosures that contain sensitive circuits. The feedthrough capacitor serves also as a connection terminal for the internal and external circuit wiring.

4.1.1 Shield-Bead Attenuation

In actual design work it is often necessary to specify the attenuation required of a bead or string of beads. At such times it is a design requirement to convert this information into an equivalent impedance. We will now examine the relationship between the two kinds of data. We will learn how to convert one type of requirement into the other by means of equations.

The insertion loss or attenuation of a bead in decibels is obtained by

$$\text{attenuation} = 20 \log_{10} \frac{E_L}{E_{LB}}$$

where E_L is the voltage across the load without the bead in the signal path, and E_{LB} is the voltage across the load after the bead has been inserted. Figure 4-7a and b show the two conditions of the equation. A desired amount of attenuation can be obtained through empirical means by simply

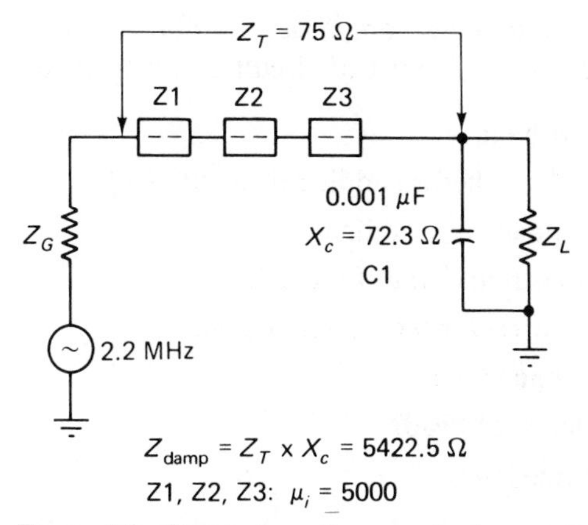

Figure 4-6 Signal absorption can be increased greatly by using a capacitor in combination with ferrite beads, as shown here. The load is in parallel with C1.

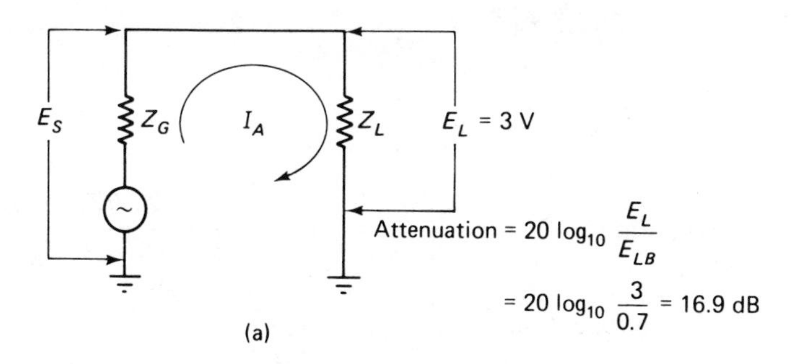

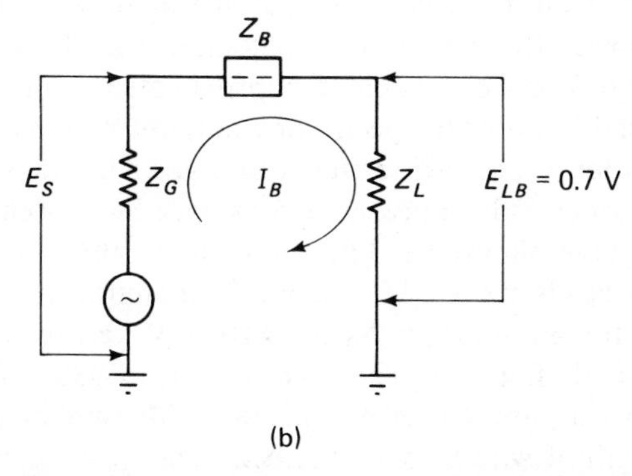

Figure 4-7 How the attenuation of a ferrite bead (Z_B) can be determined (see the text).

adding beads and making comparative voltage measurements until the design objective is met. The list of circuit designators in Fig. 4-7 is:

- E_s source voltage.
- E_L load voltage before adding a shield bead.
- E_{LB} load voltage after adding a bead.
- I_A circuit current without a bead.
- I_B circuit current after adding a bead.
- Z_G source impedance.
- Z_B impedance of bead.
- Z_L load impedance.

Attenuation in relationship to shield-bead impedance can be extracted from the following equations, which apply to the foregoing list of symbols and Fig. 4-7.

$$E_L = I_A Z_L = \frac{E_s}{Z_G + Z_L} \times Z_L$$

and

$$E_{LB} = I_B Z_L = \frac{E_s}{Z_G + Z_B + Z_L} \times Z_L$$

Also,

$$\text{attenuation} = 20 \log_{10} \frac{E_L}{E_{LB}} = 20 \log_{10} \frac{E_s Z_L/(Z_G + Z_L)}{E_s Z_L/(Z_G + Z_B + Z_L)}$$

and

$$\text{attenuation} = 20 \log_{10} \frac{Z_G + Z_B + Z_L}{Z_G + Z_L}$$

It is readily apparent from the equations that attenuation is related to the shield-bead impedance, the generator impedance, and the load impedance. It is necessary to know or be able to approximate the impedances of the generator and load in order to specify one in terms of the other.

A typical set of data for a specified ferrite bead is given in Fig. 4-8. The curves for impedance, series inductance, and resistance have been drawn versus frequency for a subminiature bead that has a μ_i of 5000. Curves for other types and sizes of beads are available from the manufacturers.

A group of curves for an Amidon Assoc. FB101-43 ferrite bead are presented in Figs. 4-9, 4-10, and 4-11. The bead has a μ_i of 950, an OD of 0.138 in., an ID of 0.051 in., and a length of 0.118 in. Measurements were made with a 0.5-in. length of wire through the bead. Figure 4-9 contains a curve for operating frequency versus impedance in ohms.

Two curves are seen in the example at Fig. 4-10. One is for resistance

Bead Dimensions

OD (in.)	ID (in.)	H (in.)	Magnetic Path Length l_e (cm)	Magnetic Cross Section A_e (cm^2)
0.138±0.008	0.051±0.004	0.128±0.010	0.64	0.0033

Magnetic Properties		
Initial permeability μ_i ±30%	5000	
Loss factor ($\tan \delta/\mu$) at 100 kHz	10×10^{-6}	
Hysteresis factor (h/μ^2)	1×10^{-5}	
Saturation flux density (B_s) at 5 O_e	4000	gauss
Remanence (B_r)	1000	gauss
Coercivity (H_c)	0.1	oersted
Resistivity	10^2	ohm-cm
Curie temperature (T_c)	>140	°C

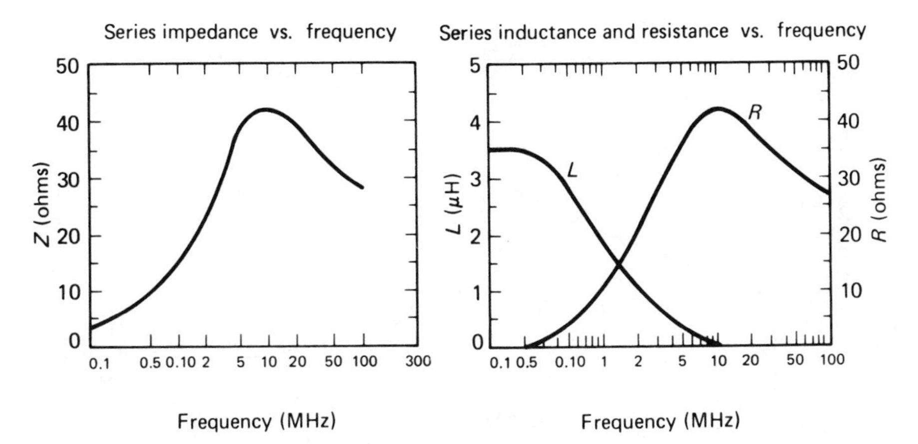

Figure 4-8 Electrical specifications for a given ferrite bead which has a μ_i of 5000. It is assumed that the bead is slipped over a straight piece of wire with pigtails no longer than ¼ in. (Courtesy of Amidon Associates.)

and the other is for inductance. Both curves are referenced to operating frequency.

The curve of Fig. 4-11 shows the relationship between attenuation and frequency. It can be seen that as the operating frequency is increased, the attenuation in decibels escalates almost linearly.

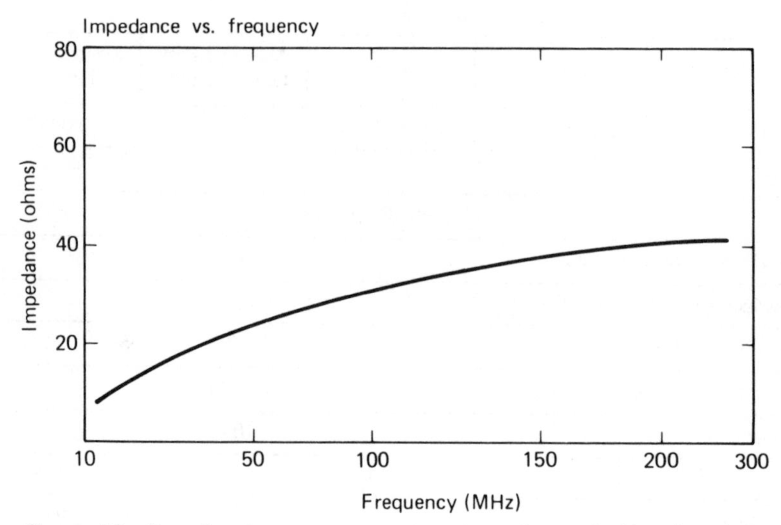

Figure 4-9 Operating frequency versus impedance for an Amidon Associates FB101-43 ferrite bead (950μ).

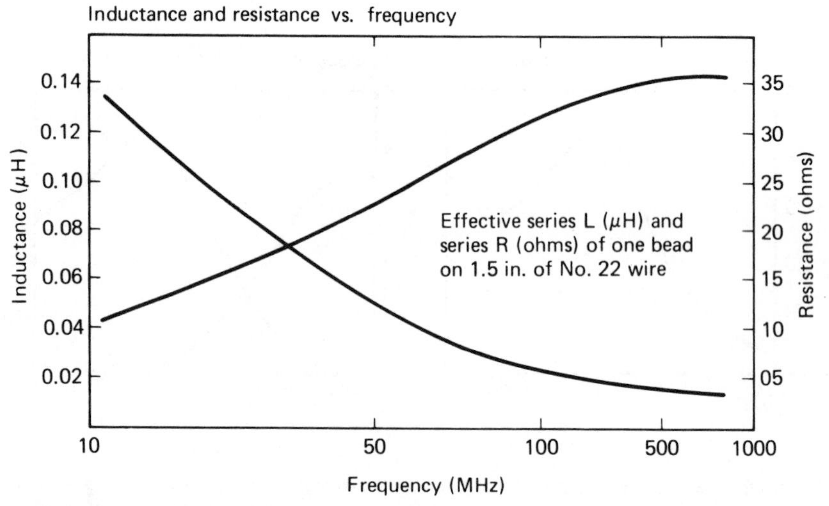

Figure 4-10 Series resistance (*R*) and inductance (*L*) versus operating frequency for the ferrite bead specified in Fig. 4-9.

4.1.2 Practical Applications for Beads

Many RF circuits are subject to self-oscillation in the VHF and UHF parts of the spectrum. The common term for this condition is "parasitic oscillation." The malady is especially common when bipolar transistors with high f_T ratings are used. The same is true of FETs that have high max-

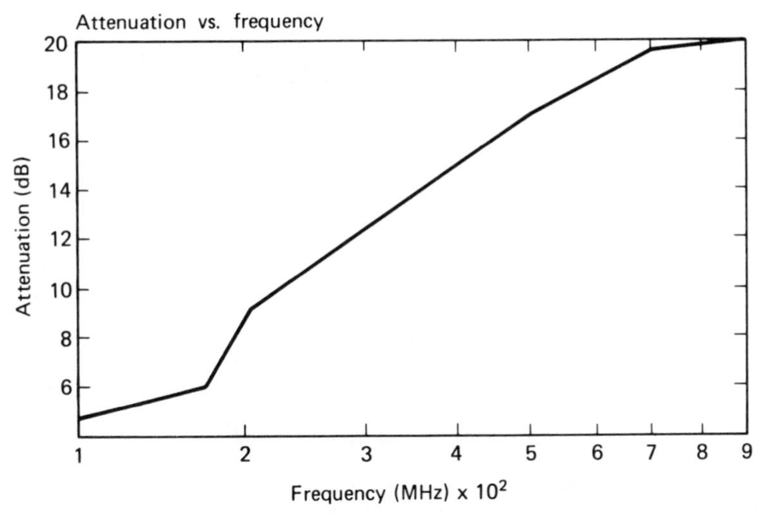

Figure 4-11 Attenuation with respect to operating frequency while using an Amidon Associates FB101-43 ferrite bead.

imum operating frequencies, such as the 2N4416 class of devices. VMOS power FETs, because of their extremely high g_{fs} ratings (250,000 μmhos being typical) and high upper-frequency limits, are prone to VHF parasitics. In all cases the anomaly is encouraged by long transistor and circuit leads. When printed circuits are used, parasitics become manifest most easily when the etched conductors are quite long between component points that relate to the active devices. Printed-circuit boards of the double-clad variety are recommended in situations where the parasitic capacitances between the ground plane and the etched foils will not impair circuit performance. This type of board material, after etching, has a solid copper surface on the component side of the board. It is connected to system ground and to the ground foils on the etched side of the board. Parasitic oscillations are discouraged because the large ground conductor helps to break up otherwise existing RF ground loops. Furthermore, small amounts of capacitance exist between the etched conductors and the ground plane. This aids in damping VHF and UHF self-oscillation.

When all other preventive measures fail to stop unwanted oscillations, insertion of one or more ferrite beads directly at the gate or base of a transistor will help. Figure 4-12 shows this technique. Z1 is a 950μ subminiature bead. It is located as close to the case of Q1 as possible. It will present negligible reactance to the 5-MHz operating frequency, but will appear as a large series impedance at VHF and higher. Similar results can be had by putting the bead in the drain or collector lead, provided that those leads are "hot" at RF. The oscillator of Fig. 4-12a has a bypassed drain, making it impractical to insert Z1 at that port. Were this circuit afflicted with VHF

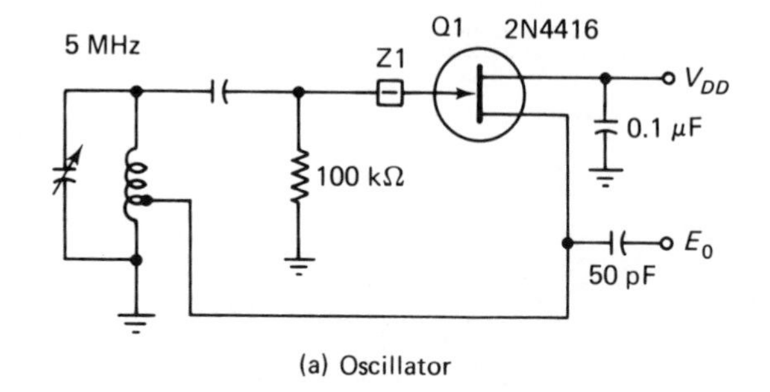

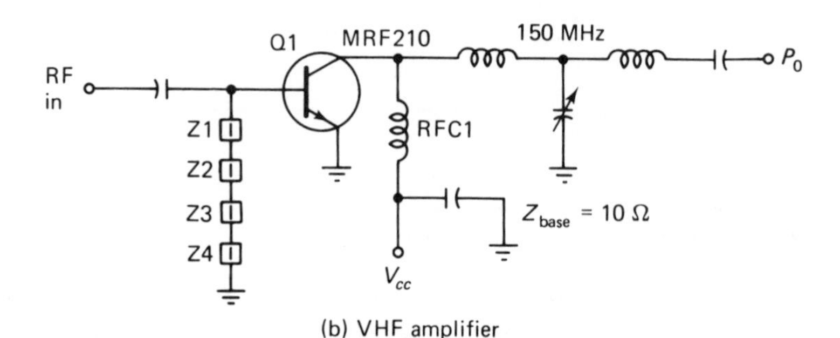

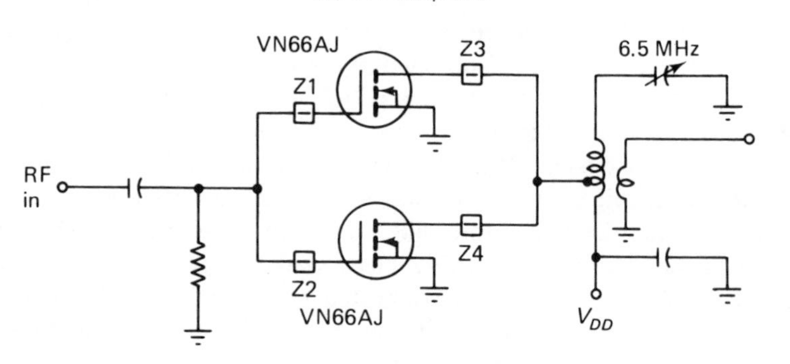

Figure 4-12 A ferrite bead is used in (a) to suppress VHF/UHF parasitic oscillations. Four beads are employed in (b) as a base impedance of low Q. Four beads are applied in (c) for parasitic suppression.

parasitics, the VHF energy would be readily apparent by observing the output waveform (E_0) by means of a wide-band oscilloscope. The VHF energy would be superimposed on the 5-MHz sine wave.

Z1 through Z4 of Fig. 4-12c are used for the purpose just discussed. But, because of the very high g_{fs} of the two VMOS power FETs, and because the devices are connected in a parallel arrangement (increased lead lengths), parasitic oscillation is a problem. Complete stability resulted in this circuit only after a bead (950 μ) was placed in both the drain and gate leads. It is practical to employ a 10- to 22-Ω ½-W noninductive resistor in place of Z1 and Z2 for parasitic suppression. In fact, many designers use that technique. However, if resistors were situated at Z3 and Z4, the dc voltage drop across them would be prohibitive: the beads offer a better solution.

In Fig. 4-12b we find Z1 through Z4 serving a different purpose. Here they provide an impedance across which the driving energy to Q1 is supplied. If we use the beads specified for the curve of Fig. 4-9, each bead will exhibit an impedance of 38 Ω at 150 MHz. A total impedance of 152 Ω will be present with the four beads—somewhat more than the minimum requirement of four times the base impedance specified (10 Ω). The primary advantage in using the beads instead of a conventional wire-wound solenoidal RF choke is that the Q will be low for Z1 through Z4 compared to a wire-wound choke. This is a desirable condition in the interest of circuit stability. Many times a solid-state amplifier will break into self-oscillation by virtue of a tuned base/tuned collector condition brought on by the RF chokes in the base and collector leads. The input and output capacitance of the transistor helps to establish the frequency of oscillation. But, if the base-return impedance is of low Q, and if the collector RF choke (RFC1) of Fig. 4-12b is of low Q, self-oscillation is unlikely. The cost of four beads versus that of a commercially made RF choke is similar.

4.1.3 Beads as Q Killers

If for the purpose of manufacturing expediency we elected to use high-Q solenoidal or layer-wound RF chokes for impedances, we might easily encounter RF instability. This possibility was discussed in Sec. 4.1.2, where we recognized the potential instability of the circuit of Fig. 4-12b. An RF power amplifier which contains two conventional RF chokes is shown in Fig. 4-13a. RFC1 and RFC2 could lead to unwanted tuned base/tuned collector instability if the self-resonant characteristics of the tuned circuits formed by the chokes and the existing circuit capacitances were similar. One solution to the problem would be to swamp one or both of the RF chokes by shunting the inductor(s) with a low-ohmage, noninductive resistor. This would lower the Q and discourage self-oscillation of Q1. The major shortcoming of using parallel resistance is that part of the amplifier power would be dissipated in the resistance: the trade-off of gain and stage efficiency for stability might be prohibitive as far as the design objective is concerned.

If we were to add high-μ ferrite beads as shown at RFC1 and RFC2 of Fig. 4-13a, the Q of the chokes would drop to a very low value. Z1 and Z2 in such an example would be miniature beads with a μ_i of 950. A single bead would be placed over the ground-return pigtail of each choke, and as close to the body of the choke as possible. Laboratory investigations, while

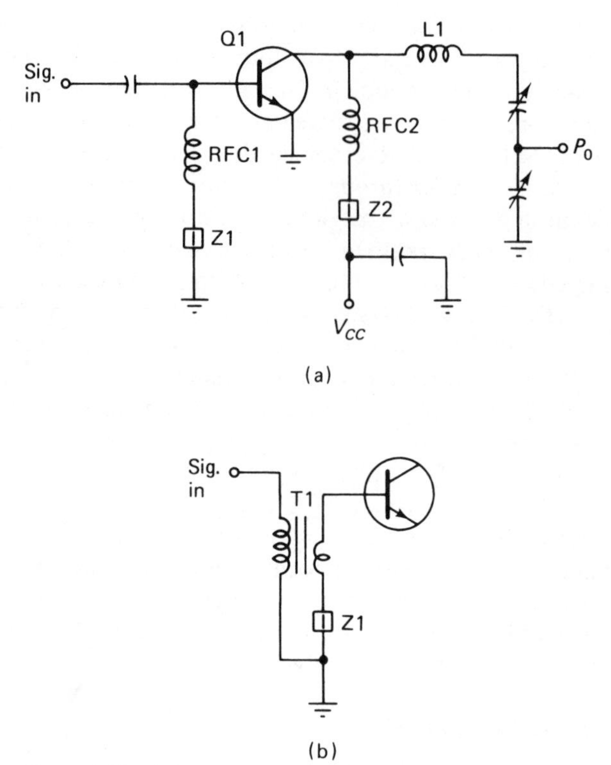

Figure 4-13 Z1 is used in (a) and (b) to lower the circuit Q (see the text) and enhance amplifier stability.

using a Boonton 160-A Q meter, revealed a Q reduction of 87.5% when a single miniature 950μ bead was added to the pigtail of a 10-μH solenoidal-wound RF choke. The measurement was made at 25 MHz. The Q_u dropped from 80 to 10, and the effective inductance increased approximately 1 μH. Next, two beads were used and the measurements repeated. The Q_u degraded to approximately 3 and the inductance rose to 12 μH—a Q reduction of some 96%.

We might ask ourselves at this juncture: Why not simply use a string of ferrite beads in place of a bead/RF choke combination? At VHF and higher this would be the more practical technique. However, in the HF part of the spectrum it might require a prohibitive number of beads to realize the impedance required for the circuit. Therefore, it is less difficult to utilize a wire-wound choke and lower the Q by adding one or more ferrite beads.

There are some circuits in which the base return of a power transistor can tolerate a small amount of series resistance without imposing a threat to the Class C transistor amplifier. In such an example we could substitute a 50- to 100-Ω composition resistor for Z1. This would lower the choke Q to an acceptable level. The method is not recommended, as the reverse-breakdown profile that resulted could damage the transistor during periods when drive was applied. The destruction would be gradual in a normal situation, and the fault would be observed as beta degradation.

The *Q*-killing procedure shown in Fig. 4-13a can be applied in the case of the broadband transformer, T1, seen in Fig. 4-13b. One or more high-permeability beads would simply be placed over the return lead of the T1 secondary winding, as shown. The bead permeability discussed in this section (950) is not mandatory. The author has learned through years of circuit design and application that a 950μ ferrite material represents a good compromise for circuits from 2 to 450 MHz. However, beads with lower μ_i characteristics can be employed provided that the desired results are realized. At the high end of the frequency range just mentioned (100 to 450 MHz), permeabilities of 125 and 250 work quite nicely for use as shown in Fig. 4-13.

4.1.4 Beads for Decoupling

At VHF and higher the ferrite bead becomes one of the most effective components available at low cost for RF filtering and decoupling. Z1 through Z6 in Fig. 4-14 are used in the V_{CC} supply lines to prevent unwanted migration of signal energy from one stage to the other. If feedback voltage was permitted to travel along the V_{CC} leads, instability could very easily result at Q1, Q2, or both. C1 and C2 prevent Q degradation of RFC1 and RFC2 which would otherwise be caused by the use of Z1/Z2 and Z5/Z6. These capacitors become part of the RF decoupling networks along with C3, C4, and Z3/Z4. Three different values of capacitance are used in the

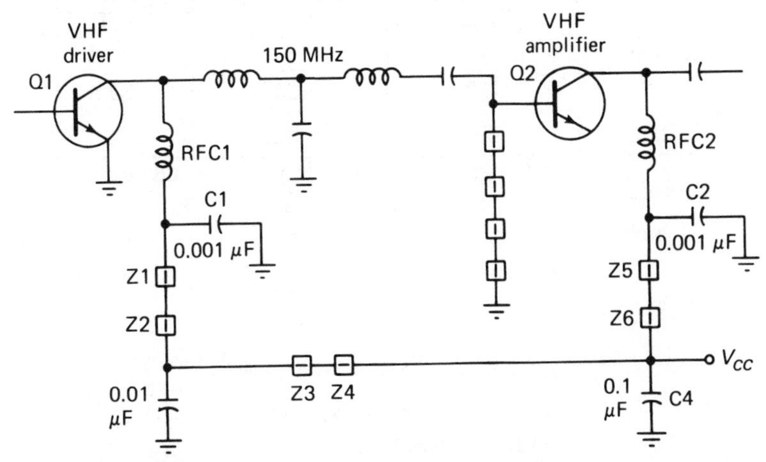

Figure 4-14 Example of ferrite beads in the decoupling networks of the V_{cc} line. Q2 uses four beads as a base-return impedance.

decoupling circuit. This provides effective bypassing over a relatively wide range of frequencies, thereby discouraging self-oscillation in and below the VHF region. Beads that have μ_i values of 125 to 950 are suitable for the circuit of Fig. 4-14. Four beads are shown between the base of Q2 and ground. They function as a base-return impedance, as is the case with the circuit of Fig. 4-12b.

4.1.5 Reducing Incidental Radiation

Certain practical and legal restrictions exist with regard to the maximum tolerable incidental radiation from specified RF generators and transmitters. A designer must ensure that equipment will not interfere with other pieces of apparatus in a composite or interfacing system. Furthermore, excessive incidental radiation from a specified module or its external wiring could easily cause interference to other services, such as TV receivers, FM receivers, or commercial two-way radio services nearby. Ferrite beads are excellent in filtering arrangements of the type illustrated in Fig. 4-15.

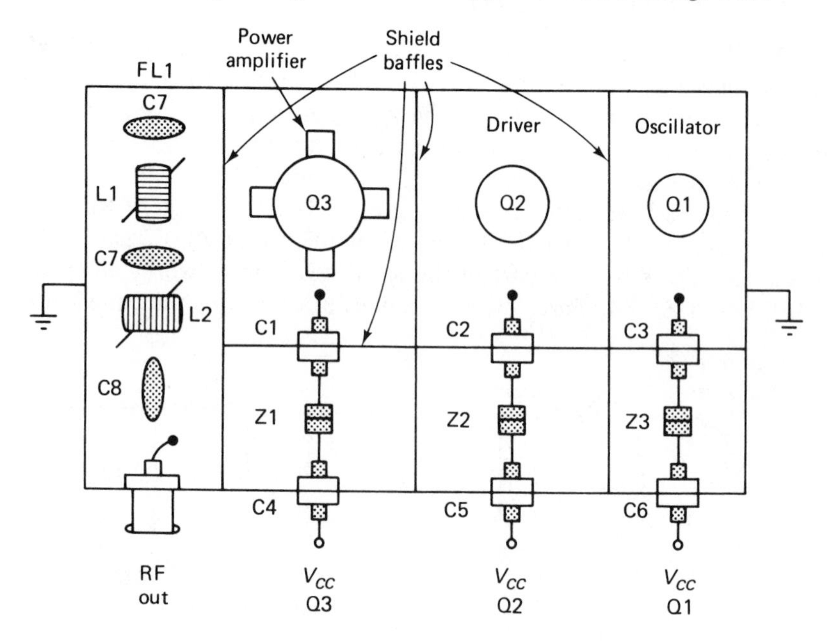

C1-C6, incl: 0.001 μF feedthrough

Z1, Z2, Z3: 2 each 950μ_i bead

FL1: Harmonic filter

Figure 4-15 Suggested method for utilizing ferrite beads to prevent incidental radiation or pickup of RF energy when a circuit module must be well shielded.

The example shows a compartmentized three-stage transmitter that is structured in a shield-compartment format. There are seven separate sections to the RF module. It is assumed that each is RF-tight with respect to the others. Additionally, the top and bottom openings of the assembled module are enclosed by means of RF-tight covers. Double-clad printed-circuit-board sections are suitable for constructing the compartments. Die-cast aluminum boxes are frequently used for the purpose as well, but must be cast with the desired compartments to assure ease of assembly.

The V_{cc} supply lines to each of the three stages in Fig. 4-15 contain effective RF filters which are composed of two feedthrough types of capacitors (C1 through C6) and two $950\mu_i$ ferrite beads (Z1, Z2, and Z3). These networks also serve to decouple the stages from one another, as was done at Fig. 4-14. The feedthrough capacitors provide a mechanical advantage: they serve also as through-terminals for the operating voltages.

If the equipment being treated for incidental radiation operates in the HF spectrum or below, Z1 through Z3 can be changed to large high-μ ferrite beads which have several turns of wire looped through them. This will provide a much higher series impedance than can be secured from the miniature beads indicated. Figure 4-4a shows the method under discussion. Another feature of the shielding and filtering procedure of Fig. 4-15 is that unwanted energy exterior to the RF module cannot enter the circuit via the supply lines. Stray energy pickup is defeated because of the shield box and its compartments.

Table 4-1 contains a listing of various sizes and styles of ferrite beads. Data on the various permeabilities are provided at the end of the table. The Class 27 beads can be obtained with wire leads for use in high-production operations. Having the leads attached permits the use of insertion mechanisms which require components to be supplied in large tape reels. Such machines are programmed to extract the components from the reels (resistors, diodes, capacitors, beads, etc.), bend the pigtails, and insert the parts in the proper pc-board holes.

The multihole beads are useful when the impedance needs to be increased over what would be presented by one or more single-hole beads. The turns of wire are passed through the various bead holes to increase the inductance. The beads with very high μ_i values listed in Table 4-1 are excellent for use at medium frequency and lower.

4.2 Ferrite Sleeves

There are a variety of applications for ferrite sleeves. Of significance to the designer is the fact that a sleeve used over an axial-lead inductor will multiply the inductance by a factor of 2 to 10, depending upon the μ_i of the material and its bulk. The basic sleeve is seen in Fig. 4-16a. The right-hand

TABLE 4-1

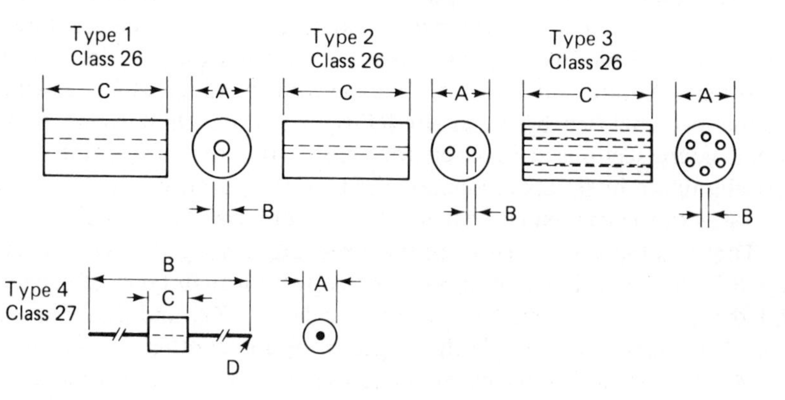

Part Number	*Material*	*Type*		*A Dim.*	*B Dim.*	*C Dim.*	*D Dim.*
2673004501	**73**	**1**	**in**	**0.037/0.039**	**0.019/0.022**	**0.192/0.198**	
			mm	0.94/0.99	0.48/0.56	4.88/5.03	
2673004601	**73**	**1**	**in.**	**0.040/0.043**	**0.025/0.029**	**0.155/0.163**	
			mm	1.02/1.09	0.64/0.74	3.94/4.14	
2673004701	**73**	**1**	**in.**	**0.052/0.056**	**0.027/0.031**	**0.085/0.095**	
			mm	1.32/1.42	0.69/0.79	2.16/2.41	
2643004801	**43**	**1**	**in.**	**0.076/0.084**	**0.033/0.038**	**0.100/0.112**	
			mm	1.93/2.13	0.84/0.97	2.54/2.84	
2643002201	**43**	**1**	**in.**	**0.068/0.076**	**0.041/0.045**	**0.400/0.430**	
			mm	1.73/1.93	1.04/1.14	10.16/10.92	
2673002201	**73**	**1**	**in.**	**0.068/0.076**	**0.041/0.045**	**0.400/0.430**	
			mm	1.73/1.93	1.04/1.14	10.16/10.92	
2673000501	**73**	**1**	**in.**	**0.074/0.078**	**0.041/0.045**	**0.053/0.067**	
			mm	1.88/1.98	1.04/1.14	1.35/1.70	
2643000201	**43**	**1**	**in.**	**0.074/0.078**	**0.041/0.045**	**0.140/0.160**	
			mm	1.88/1.98	1.04/1.14	3.56/4.06	
2673000201	**73**	**1**	**in.**	**0.074/0.078**	**0.041/0.045**	**0.140/0.160**	
			mm	1.88/1.98	1.04/1.14	3.56/4.06	
2673004901	**73**	**1**	**in.**	**0.107/0.117**	**0.065/0.071**	**0.400/0.420**	
			mm	2.71/2.97	1.65/1.80	10.16/10.67	
2643004101	**43**	**1**	**in.**	**0.128/0.148**	**0.029/0.033**	**0.160/0.190**	
			mm	3.25/3.76	0.74/0.84	4.06/4.83	
2643004201	**43**	**1**	**in.**	**0.128/0.148**	**0.029/0.033**	**0.330/0.370**	
			mm	3.25/3.76	0.74/0.84	8.38/9.40	
2643000401	**43**	**1**	**in.**	**0.130/0.146**	**0.047/0.055**	**0.045/0.055**	
			mm	3.30/3.71	1.19/1.40	1.14/1.40	
2643000101	**43**	**1**	**in.**	**0.130/0.146**	**0.047/0.055**	**0.110/0.126**	
			mm	3.30/3.71	1.19/1.40	2.79/3.20	
2664000101	**64**	**1**	**in.**	**0.130/0.146**	**0.047/0.055**	**0.110/0.126**	
			mm	3.30/3.71	1.19/1.40	2.79/3.20	
2673000101	**73**	**1**	**in.**	**0.130/0.146**	**0.047/0.055**	**0.110/0.126**	
			mm	3.30/3.71	1.19/1.40	2.79/3.20	
2675000101	**75**	**1**	**in.**	**0.130/0.146**	**0.047/0.055**	**0.226/0.246**	
			mm	3.30/3.71	1.19/1.40	5.74/6.25	
2643000601	**43**	**1**	**in.**	**0.130/0.146**	**0.047/0.055**	**0.150/0.170**	
			mm	3.30/3.71	1.19/1.40	3.81/4.32	
2673000601	**73**	**1**	**in.**	**0.130/0.146**	**0.047/0.055**	**0.150/0.170**	
			mm	3.30/3.71	1.19/1.40	3.81/4.32	
2643000301	**43**	**1**	**in.**	**0.130/0.146**	**0.047/0.055**	**0.226/0.246**	
			mm	3.30/3.71	1.19/1.40	5.74/6.25	
2664000301	**64**	**1**	**in.**	**0.130/0.146**	**0.047/0.055**	**0.226/0.246**	
			mm	3.30/3.71	1.19/1.40	5.74/6.25	

2673000301	**73**	**1**	**in.**	**0.130/0.146**		**0.047/0.055**	**0.226/0.246**	
			mm	3.30/3.71		1.19/1.40	5.74/6.25	
2643000701	**43**	**1**	**in.**	**0.130**		**0.047/0.55**	**0.485/0.515**	
			mm	3.30/3.71		1.19/1.40	12.32/13.08	
2664000701	**64**	**1**	**in.**	**0.130/0.146**		**0.047/0.055**	**0.485/0.515**	
			mm	3.30/3.71		1.19/1.40	12.32/13.08	
2673000701	**73**	**1**	**in.**	**0.130/0.146**		**0.047/0.055**	**0.485/0.515**	
			mm	3.30/3.71		1.19/1.40	12.32/13.08	
2643001301	**43**	**1**	**in.**	**0.135/0.145**		**0.065/0.075**	**0.224/0.244**	
			mm	3.43/3.68		1.65/1.91	5.69/6.20	
2643001401	**43**	**1**	**in.**	**0.153/0.169**		**0.074/0.084**	**0.457/0.487**	
			mm	3.89/4.29		1.88/2.13	11.61/12.37	
2663001201	**63**	**1**	**in.**	**0.157/0.181**		**0.063/0.079**	**0.185/0.209**	
			mm	3.99/4.60		1.60/2.01	4.70/5.31	
2633022401	**33**	**1**	**in.**	**0.190/0.210**		**0.057/0.067**	**0.240/0.260**	
			mm	4.83/5.33		1.45/1.70	6.10/6.60	
2643022401	**43**	**1**	**in.**	**0.190/0.210**		**0.057/0.67**	**0.240/0.260**	
			mm	4.83/5.33		1.45/1.70	6.10/6.60	
2643021801	**43**	**1**	**in.**	**0.190/0.210**		**0.057/0.067**	**0.422/0.452**	
			mm	4.83/5.33		1.45/1.70	10.72/11.48	
2664021801	**64**	**1**	**in.**	**0.190/0.210**		**0.057/0.067**	**0.422/0.452**	
			mm	4.83/5.33		1.45/1.70	10.72/11.48	
2673021801	**73**	**1**	**in.**	**0.190/0.210**		**0.057/0.067**	**0.422/0.452**	
			mm	4.83/5.33		1.45/1.70	10.72/11.48	
2664001101	**64**	**2**	**in.**	**0.208/0.232**	(2)	**0.029/0.041**	**0.456/0.488**	
			mm	5.28/5.89	(2)	0.74/1.04	11.58/12.40	
2643225111	**43**	**3**	**in.**	**0.226/0.246**	(6)	**0.035/0.041**	**0.384/0.404**	
			mm	5.74/6.25	(6)	0.89/1.04	9.75/10.26	
2664225111	**64**	**3**	**in.**	**0.226/0.246**	(6)	**0.035/0.041**	**0.384/0.404**	
			mm	5.74/6.25(6)		0.89/1.04	9.75/10.26	
2664000901	**64**	**2**	**in.**	**0.240/0.260**	(6)	**0.045/0.055**	**0.456/0.486**	
			mm	6.10/6.60	(2)	1.14/1.40	11.58/12.34	
2643000801	**43**	**1**	**in.**	**0.291/0.301**		**0.089/0.099**	**0.287/0.307**	
			mm	7.39/7.65		2.26/2.51	7.29/7.80	
26640008010	**64**	**1**	**in.**	**0.291/0.301**		**0.089/0.099**	**0.287/0.307**	
			mm	7.39/7.65		2.26/2.51	7.29/7.80	
2673000801	**73**	**1**	**in.**	**0.291/0.301**		**0.089/0.099**	**0.287/0.307**	
			mm	7.39/7.65		2.26/2.51	7.29/7.80	
2643006801	**43**	**1**	**in.**	**0.365/0.385**		**0.187/0.200**	**0.400/0.420**	
			mm	0.27/9.78		4.75/5.08	10.16/10.67	
2677006801	**77**	**1**	**in.**	**0.365/0.385**		**0.187/0.200**	**0.400/0.420**	
			mm	9.27/9.78		4.75/5.08	10.16/10.67	
2643002401	**43**	**1**	**in.**	**0.370/0.390**		**0.190/0.205**	**0.180/0.200**	
			mm	9.40/9.91		4.83/5.21	4.57/5.08	
264340001	**43**	**1**	**in.**	**0.542/0.582**		**0.240/0.260**	**1.095/1.155**	
			mm	13.77/14.78		6.10/6.60	27.81/29.34	
2743001111	**43**	**4**	**in.**	**0.128/0.148**		**2.720/2.780**	**0.160/0.190**	**=22 AWG TCW***
			mm	3.25/3.76		69.09/70.61	4.06/4.83	× 1½ (38.1mm)
2743002121	**43**	**4**	**in.**	**0.128/0.148**		**2.720/2.780**	**0.330/0.370**	**=22 AWG TCW***
			mm	3.25/3.76		69.09/70.61	8.38/9.40	× 1½ (38.1mm)

Available bead sizes and permeabilities. Courtesy Fair-Rite Corp. Initial permeabilities are given here with material type: 33 = 800 mu; 43 = 850 mu; 63 = 40 mu; 64 = 250 mu; 73 = 2500 mu; 75 = 5000 mu; 77 = 1800 mu;

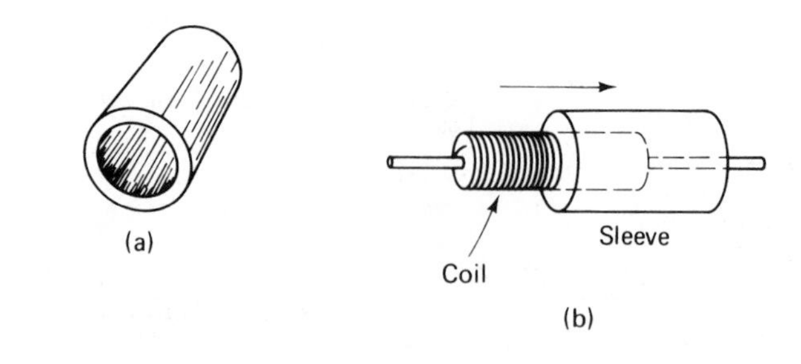

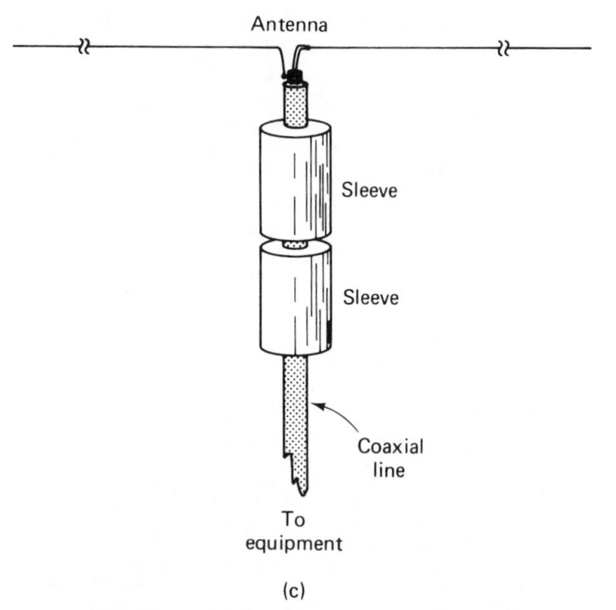

Figure 4-16 The basic ferrite sleeve is shown pictorially in (a). A solenoidal inductor is shown partially inserted into the sleeve in (b). Ferrite sleeves can be used to decouple coaxial RF transmission lines as shown in (c).

illustration in part (b) shows how the sleeve and a solenoidal inductor can be used in combination to elevate the coil inductance.

Additional to the feature of increased inductance, the sleeve functions as an effective shield for the coil. Still another advantage in using the sleeve is that it offers physical protection to the inductor. This eliminates the need for a special shield can or the added burden of encapsulation. The wall thickness of ferrite sleeves is 12% of the sleeve outside diameter, or greater. This is a manufacturing requirement that makes it practical to fabricate the sleeves by means of the compaction-pressing or extrusion techniques. The net result is beneficial to designers who choose to use the sleeve as a protective covering for the coil with which it is used.

Sleeves are available from Fair-Rite Corp., for example, in permeabilities from 40 to 1200. The available sizes range from 0.115 in. OD/0.155 in. long to 1.025 in. OD/2.03 in. long.

Another practical use for the ferrite sleeve is seen in Fig. 4-16c. They can be used over a coaxial transmission line to serve as decoupling sleeves. This helps to prevent RF energy from traveling on the outer conductor of the line. This technique is especially useful when arrays of gain antennas are combined. The decoupling sleeves discourage unwanted radiation of the harnessing sections of the array. Such radiation can destroy the array pattern and gain if it is great enough in magnitude.

Similarly, ferrite sleeves can be used in RFI/EMI suppression by placing them over signal cables which enter and leave the various parts of an electronics system. The sleeve can be affixed at the desired location on the cable by means of Silastic compound or epoxy cement. Heat-shrink tubing is also excellent for securing the sleeve to the cable.

4.3 Ferrite Balun Cores

The ferrite balun core of Fig. 4-5c is an excellent substitute for a group of toroid cores that have been combined for broadband transformer applications (Fig. 3-20). The practical limit is set by the sizes of available balun cores. Most of them are relatively small in cross-sectional area. This places a restriction on the amount of power that can be handled safely with ferrite baluns.

Although the term "balun" refers specifically to a *balanced-to-unbalanced* transformation of impedance levels, there is no reason why balun cores cannot be used for all manner of conventional or transmission-line transformers. Figure 4-17 shows a 150-W solid-state linear amplifier for use from 2 to 30 MHz. The input transformer uses a balun core that is ⅞ in. long and ½ in. wide. Driving power for the amplifier is on the order of 6 W. The larger transformer on the module is used at the amplifier output. It is fashioned from two 1⅞-in. lengths of ferrite tubing. Each is ½ in. in diameter. End plates have been made from printed-circuit board as shown in Fig. 3-20. However, the principle of operation for the two transformers seen in Fig. 4-17 is the same. Both are conventional transformers.

A pictorial view of how we might employ a balun core for a broadband conventional transformer is given in Fig. 4-18. The example in part (a) shows the core from one end. The shield braid from a small piece of coaxial cable has been placed in the core holes, then fanned as shown. The braid is passed through the holes so that fanning can be done at each end of the core. Solder is flowed over the fanned areas to secure the braid in the desired position. At the far end of the core in Fig. 4-18, the two sections of braid are joined and soldered. This results in the winding C–D–E of Fig.

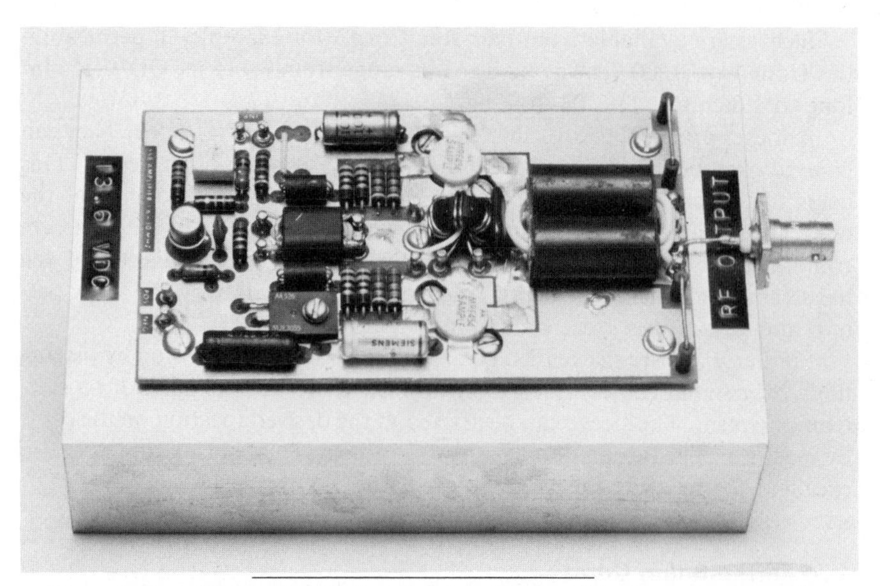

Figure 4-17 A 150-W solid-state broadband amplifier in which a balun core is used for the input transformer. The larger transformer (output) is a conventional type which contains two long cylinders of ferrite. This amplifier was designed by Helge Granberg of Motorola Semiconductor Products, Inc.

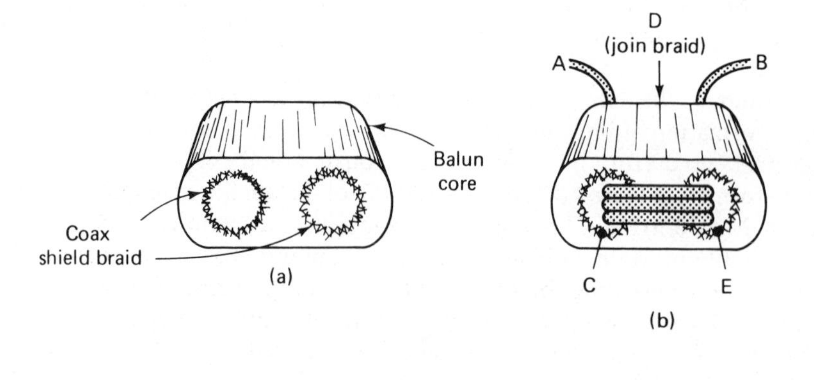

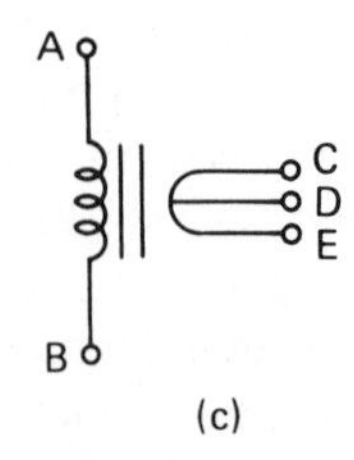

Figure 4-18 Method of using coaxial-cable shield braid as the CDE winding of (c). The braid is passed through the holes of a balun core as shown in (a) and (b).

4-18c. In effect, we have a single-turn conductor through the core, with D being the center tap. The same result can be obtained by using thin-wall lengths of brass tubing in place of the shield braid.

Winding A–B is looped through the core after the process shown at A is completed. Winding C–D–E becomes the basis for the turns ratio of the transformer, since it is fixed at one turn. Thus, if a 10:1 impedance ratio was required, winding A–B would consist of 3.16 turns of insulated wire. In a practical situation we would use a three-turn winding for A–B and tolerate the slight mismatch that would result.

In utilizing this type of transformer construction, we need to pay attention to the rule established in Chapter 3: the transformer windings must exhibit an inductive reactance that is at least four times the characteristic impedance of the circuit to which the transformer winding connects. Therefore, if T1 of Fig. 4-18 were used to match a single-ended 50-Ω driving source to a 5-Ω base-to-base impedance in a push-pull solid-state amplifier, winding C–D–E would need to have an X_L of 20 Ω or greater at the lowest operating frequency of the amplifier. Assuming in this case that the lowest frequency was 2 MHz, winding C–D–E would require 1.59 μH of inductance, or greater. The balun core would be chosen for a permeability that ensured the design objective.

Ferrite balun cores are used widely at the input ports of TV and FM tuners for the purpose of converting balanced 300-Ω feed line to the single-ended 75-Ω receiver input terminal. The cable-TV industry employs numerous balun cores for similar applications.

4.4 Pot Cores

A typical pot core is shown before and after winding and assembly in Fig. 4-19. This device is excellent when the designer requires a large amount of inductance and a high degree of magnetic self-shielding. For example, we could fabricate a 4-H inductor on a pot core that had a diameter of only $\frac{9}{16}$ in. Furthermore, excellent values of Q can be obtained when using pot cores.

Another advantage in the use of pot cores is the reduction in manufacturing cost with respect to winding equipment and winding time. Pot cores contain plastic bobbins on which the inductor or transformer windings are placed. The winding procedure is far less complex than when toroid cores are used for the same purpose.

Owing to the small size of pot cores, miniaturization becomes a routine procedure when an inductor or transformer is required in a circuit. The mounting techniques for pot cores make them compatible with etched-circuit boards and metal chassis. An assembled pot core can be impregnated to protect the winding from dirt, moisture, and fungus. The impregnation

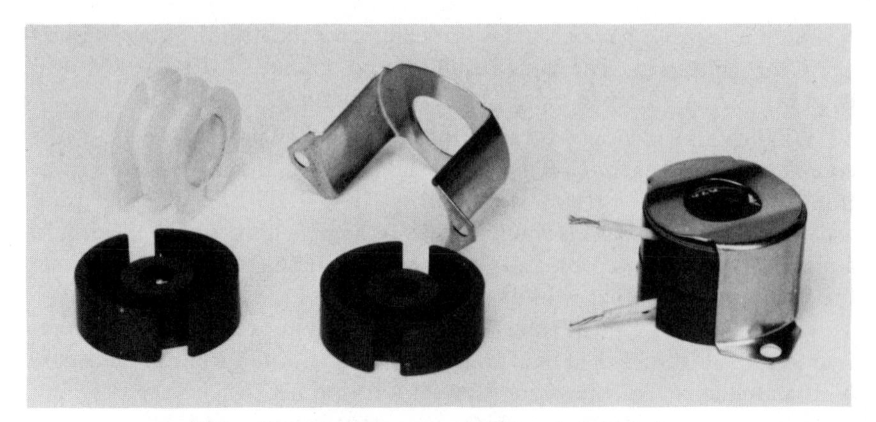

Figure 4-19 Typical pot core before and after winding.

process will also help to keep the core halves tensioned properly. Any reduction of pressure will cause the inductance of the pot-core winding to decrease. Similarly, dirt or grease on the mating surfaces of the core halves must be avoided to ensure maximum inductance. It is important, therefore, to prevent varnish or encapsulating material from appearing on the mating surfaces of the core. Even the thinnest of foreign-matter layers will reduce the core A_L.

4.4.1 Pot-Core Hardware

Two common styles of pot core are depicted in Fig. 4-20. The assembly seen in part (a) uses a metal mounting clamp to hold the core halves tightly together. The base plate and the metal clamp are held together by means of screws and nuts, thereby confining the pot core between them. Alternatively, the entire assembly can be snugged against a circuit board or metal chassis. One or more shims (washers) can be used at the bottom of the "sandwich" to provide the desired tension. Slots in the core halves permit bringing the transformer or inductor leads out of the assembly.

The inductance adjustor screw at the top of Fig. 4-20a is an optional feature of some pot cores. It can be used to vary the core-winding inductance by approximately $\pm 13\%$. This screw contains a ferrite slug that bridges the air gap within the core to increase the inductance. This feature is especially useful when precise final alignment of high-Q filter sections is required.

The core assembly shown in Fig. 4-20b is entirely suitable for most noncritical circuit applications. The winding inductance is predetermined within reasonable limits of tolerance by means of the A_L factor which relates to the core being used. A nylon nut and screw is used to hold the core halves together and to affix the completed assembly to the chassis or pc board.

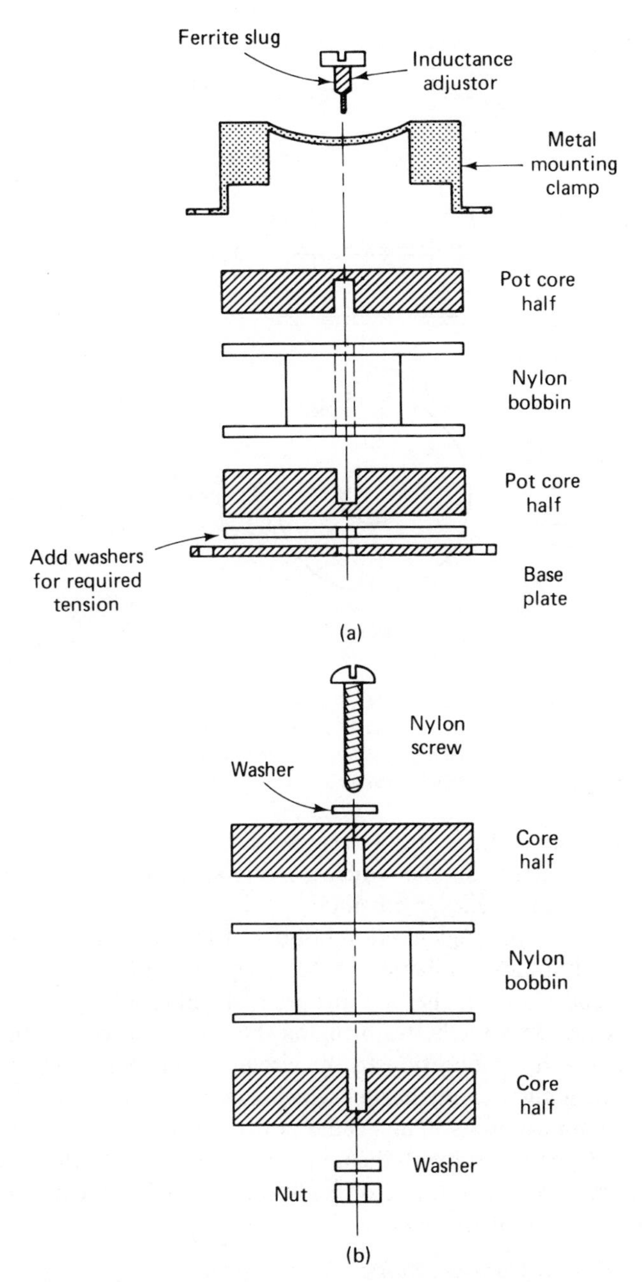

Figure 4-20 Breakaway view of a pot-core assembly which has a fine-adjustment screw and metal mounting clamp (a). The assembling in (b) mounts by means of a long nylon screw and hew nut.

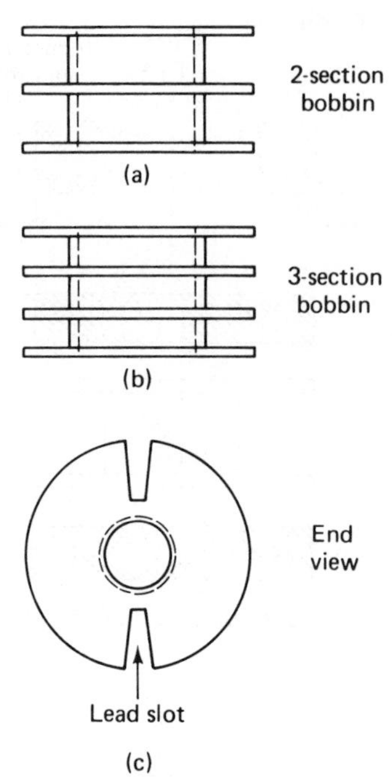

Figure 4-21 Pot-core bobbins are available with 2 (a) or more (b) winding sections. An end view of a bobbin is seen in (c).

Although ferrous or brass nuts and screws can be used for the purpose, they will have some effect on the winding inductance and Q. Therefore, they should be avoided whenever possible.

In addition to the single-section bobbin of Fig. 4-20, there are two- and three-section bobbins available. These are seen in Fig. 4-21a and b. The slots in the end plates of the bobbins are situated to align with the slots in the core halves. This facilitates bringing the transformer or inductor leads out of the core for connection to the circuit. Some manufacturers of pot cores provide a glass-epoxy header which can be affixed to the top or bottom of the core assembly. The header is fitted with the desired number of terminal posts, consistent with the number of leads coming through the core slots. The transformer or inductor leads are soldered to the header posts to reduce the fragility of the leads.

4.4.2 Design Considerations

When pot cores are employed in narrow-band LC types of circuits, there are some basic considerations which the designer must address prior to selecting a core:

1. The operating frequency.
2. Inductor Q at the chosen operating frequency.
3. Operational flux density (B_{op}).
4. Required inductance.
5. Temperature coefficient of the inductor.
6. Long-term stability.
7. Available space for the pot-core assembly.

These considerations are affected significantly by the *relative loss factor,* which is expressed mathematically as $1/\mu_i Q$. The core is chosen for the lowest relative loss at the operating frequency, or range of operating frequencies. This will help to ensure the highest Q that can be expected. The foregoing factor reflects the relative losses in the core. It varies with operating frequency and the actual core material being used.

Curves are available from the various manufacturers which show the characteristics of the different types of ferrite with regard to operating Q versus frequency. There are cross-over points where different materials may actually coincide, but the choice of ferrite will have to be based on all of the considerations in the seven-point criterion given earlier.

As is the case with other forms of ferrite and powdered-iron materials, the number of turns for a given value of inductance can be determined by $N = 10^3 \sqrt{L/A_L}$ or $N = 100 \sqrt{L/A_L}$, where N is the required number of turns, L is in mH, and A_L is the manufacturer's inductance index for the core material being used. An inductance/A_L/ turns nomograph is given in Fig. 4-22.

The A_L values for pot cores are standard, just as they are for toroids. For this reason they can be specified for each model of core the manufacturer produces. A variety of A_L values can be obtained from a given ferrite mix by carefully grinding a defined gap between the inner centerposts of the pot cores. Generally, the manufacturer grinds only one of the centerposts when small gaps are desired. Both posts are milled down when larger gaps are provided.

Quality factor (Q), as was stated earlier in the book, is a measure of the effects of the collective losses with respect to circuit performance. The three major loss factors are copper losses (winding), core losses, and capacitive losses in the winding. The Q is affected, therefore, by the number of coil turns and the manner in which they are wound on the pot-core bobbin (capacitive effects). The gauge and type of wire used is significant in designing for high Q. The larger wire diameters contribute to higher Q through reduced series resistance. Litz wire is considered superior to solid wire at the higher operating frequencies. This is because litz wire will reduce the eddy-

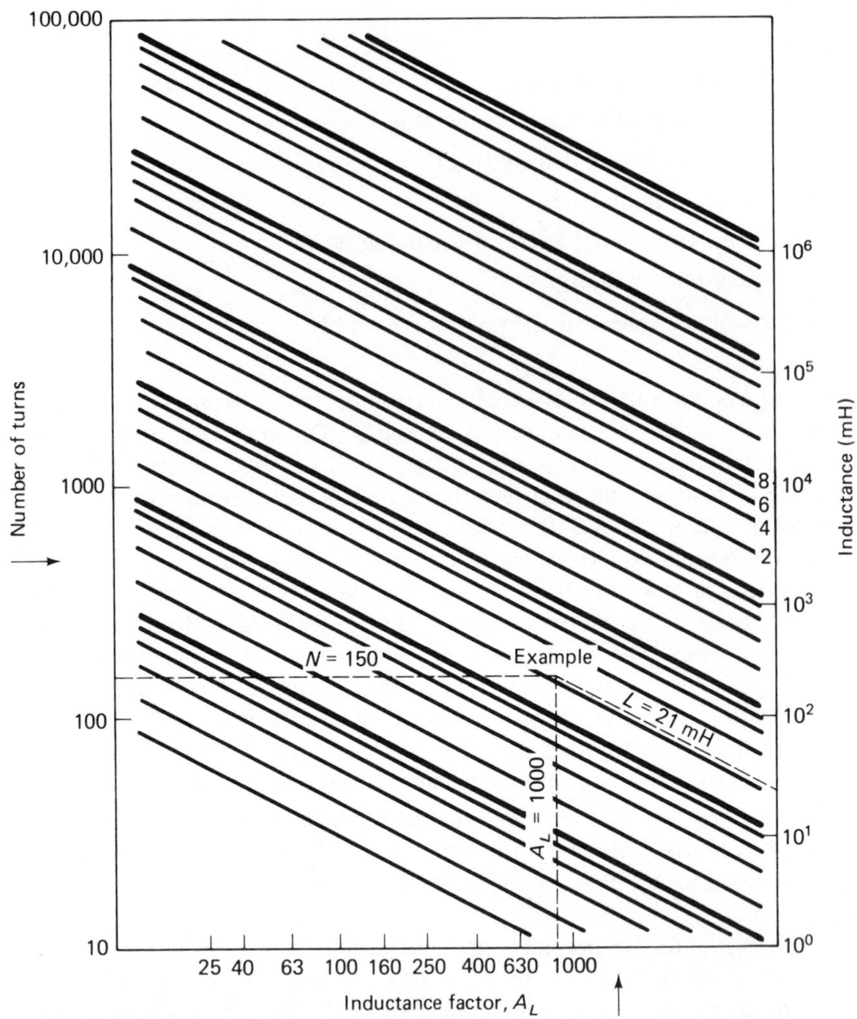

Figure 4-22 Convenient nomograph for determining inductance or wire turns when the A_L of pot core is known. (Courtesy of Magnetics Division of Spang Industries, Inc.)

current losses in the winding—an aid to higher Q. Litz-wire data are provided in Table 4-2.

As an aid to predetermining the size of bobbin needed for a given number of wire turns, it is convenient to refer to the nomograph presented in Fig. 4-23. First, the area occupied by the turns in terms of square inches or square centimeters must be determined. This will be based on the wire gauge, inclusive of the insulating material on the wire. It is also assumed that the windings will be machine-wound in layers rather than scramble-wound by hand. Tables 4-2 and 4-3 can be consulted to learn what bobbin area will be occupied by the chosen wire and number of turns. This information can then be used to select a bobbin of appropriate area by consulting Fig. 4-23.

TABLE 4-2 Litz wire data.

Litz Wire Size	TURNS *per in²*	TURNS *per cm²*	*Litz Wire Size*	TURNS *per in²*	TURNS *per cm²*
5/44	28,000	4,341	72/44	1,500	232
6/44	25,000	3,876	80/44	1,400	217
7/44	22,000	3,410	90/44	1,200	186
12/44	13,000	2,016	100/44	1,100	170
20/44	7,400	1,147	120/44	900	140
30/44	4,000	620	150/44	700	108
40/44	3,000	465	180/44	500	77
50/44	2,300	356	360/44	250	38
60/44	1,900	294			

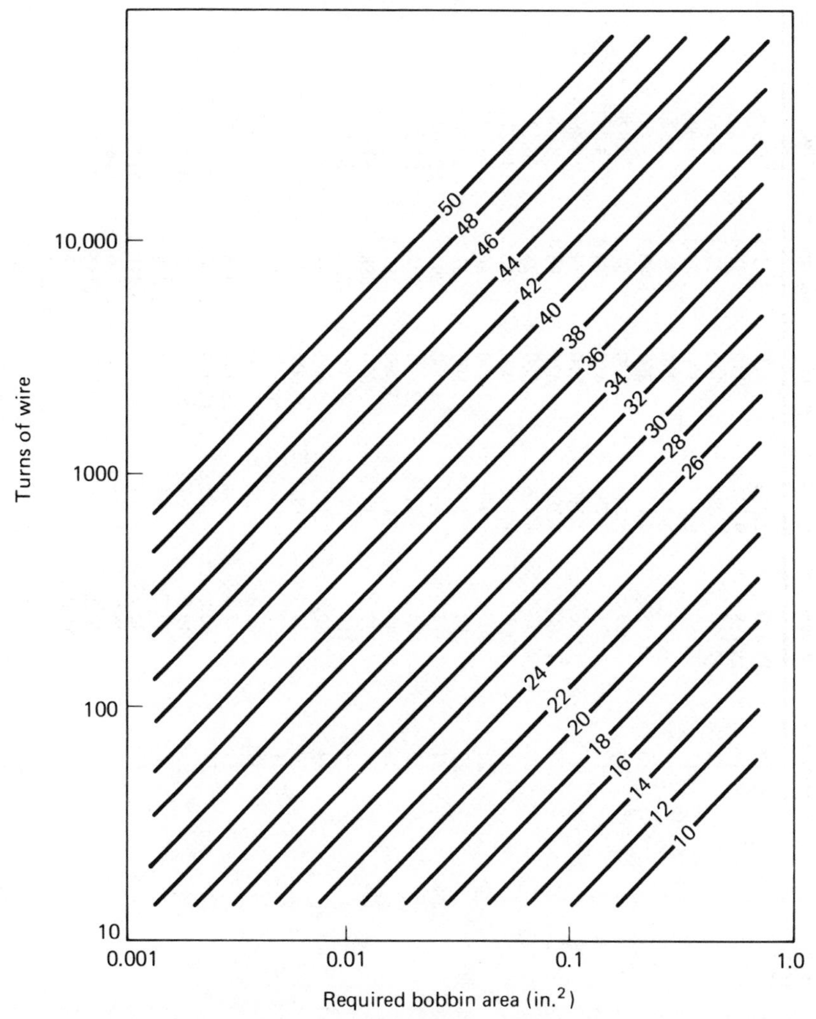

Figure 4-23 Nomograph that shows wire gauge, number of turns and bobbin area in inches². (Courtesy of Magnetics Division of Spang Industries, Inc.)

TABLE 4-3 Bobbin data.

	Maximum		Turns			
AWG Wire Size	*Wire Area Heavy*	*(cir mils) Quad*	*per in.²*	*per cm²*	*Resistance (Ω/1000 ft)*	*Current Capacity (mA) at 750 cir mil/A*
10	11,470	12,230	89	13.8	0.9987	13,840
11	9,158	9,821	112	17.4	1.261	10,968
12	7,310	7,885	140	21.7	1.588	8,705
13	5,852	6,336	176	27.3	2.001	6,912
14	4,679	5,112	220	34.1	2.524	5,479
15	3,758	4,147	260	40.3	3.181	4,347
16	3,003	3,329	330	51.2	4.020	3,441
17	2,421	2,704	410	63.6	5.054	2,736
18	1,936	2,190	510	79.1	6.386	2,165
19	1,560	1,781	635	98.4	8.046	1,719
20	1,246	1,436	800	124	10.13	1,365
21	1,005	1,170	1,000	155	12.77	1,083
22	807	949	1,200	186	16.20	853
23	650	778	1,500	232	20.30	681
24	524	635	1,900	294	25.67	539
25	424	520	2,400	372	32.37	427
26	342	424	3,000	465	41.0	338
27	272	342	3,600	558	51.4	259
28	219	276	4,700	728	65.3	212
29	180	231	5,600	868	81.2	171
30	144	188	7,000	1,085	104	133
31	117	154	8,500	1,317	131	106
32	96.0	128	10,500	1,628	162	85
33	77.4	104	13,000	2,015	206	67
34	60.8	82.8	16,000	2,480	261	53
35	49.0	67.2	20,000	3,100	331	42
36	39.7	54.8	25,000	3,876	415	33
37	32.5	44.9	32,000	4,961	512	27
38	26.0	36.0	37,000	5,736	648	21
39	20.2	28.1	50,000	7,752	847	16
40	16.0	22.1	65,000	10,077	1,080	13
41	13.0		80,000	12,403	1,320	11
42	10.2		100,000	15,504	1,660	8.5
43	8.4		125,000	19,380	2,140	6.5
44	7.3		150,000	23,256	2,590	5.5
45	5.3		185,000	28,682	3,348	4.1

Long-term stability (DF_e) of the core assembly is a matter of more than casual concern when magnetic core material is used for narrow-band applications. Stability drift causes a *decrease* in inductance. In common engineering terminology, this phenomenon is referred to as "disaccommodation."

Disaccommodation can be calculated for each pot-core size and A_L factor. The drift takes place at a logarithmic rate and the long-term shift in inductance can be determined from

$$\frac{\Delta L}{L} = DF_e \times \log \frac{T2}{T1}$$

where $\Delta L/L$ is the decrease in inductance between times T1 and T2, DF_e is the effective disaccommodation coefficient of the chosen core, and T1 and T2 are time differences. For the purpose of definition concerning the foregoing equation, T1 is the time lapse between manufacture of the core (these data are usually marked on the shipping carton) and the time it is soldered into the circuit. T2, on the other hand, is the time lapse from manufacture to the period when the product in which the core is used has ended.

Disaccommodation commences the moment the core is fabricated by the manufacturer, through its period of cooling (Curie temperature). Subsequent to this time frame, if the core is thermally or mechanically shocked, or if it is demagnetized, the inductance can increse to its original value. If this occurs, disaccommodation starts anew. In view of these traits, the designer must take all the possibilities into account when designing a circuit. But if a relatively normal environment is expected for the equipment, the long-term changes in pot-core inductance will be small. Most of the change will take place during the first few months after the core is manufactured. It is recommended, therefore, that the unwound cores be allowed to age a sufficient length of time before they are placed in the manufactured product.

Effective permeability (μ_e) is as important when working with pot cores as it is with any magnetic-core material. Unlike the toroid core, a pot core has a defined gap. The air gap determines the μ_e and hence the A_L factor for any given type of pot core. The larger the air gap, the lower the A_L. Conversely, the larger the air gap, the better the stability of the inductance with time and temperature. A definite degradation of the Q will accompany increases in air gap. These conditions do, therefore, conflict somewhat with one another from a design-objective viewpoint. Since stability and Q are both highly desirable in narrow-band circuit work, the air gap must be a compromise value in order to strike an acceptable balance between the two desired conditions—Q and stability.

A representative circuit for a pot core with an air gap is seen in Fig. 4-24.

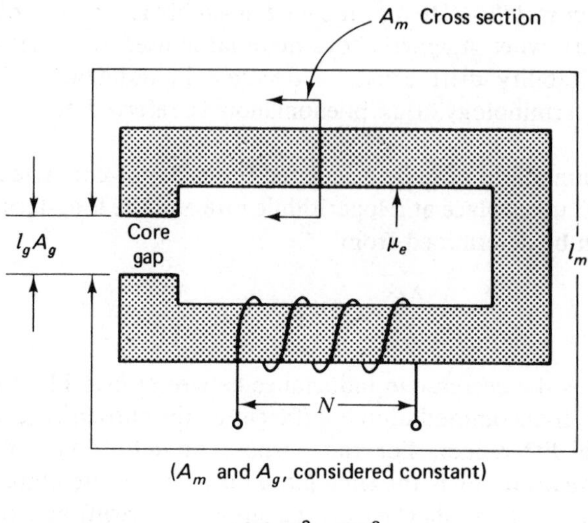

Figure 4-24 Representation of a pot core with its gap. Design considerations are detailed in the text.

For an inductor wound in the manner shown, the inductance can be found by

$$L = \frac{0.4\pi N^2 \times 10^{-8}}{\dfrac{l_m}{\mu_i A_m} + \dfrac{l_g}{A_g}} \qquad \text{henries}$$

where A_m and A_g are assumed constant, A_m is the effective magnetic path area in cm², A_g the effective gap area in cm², N the number of coil turns, l_m the length of the magnetic path in centimeters, l_g the length of the air gap, and μ_i the initial permeability of the core. The denominator is written customarily as

$$\Sigma \frac{l}{\mu_e A}$$

which is the reluctance of the magnetic circuit.

We can avoid thought complication by regarding the pot core as a device with a closed, homogeneous magnetic path which has a μ_e that enables us to determine that

$$L = \mu_e \left(\frac{0.4\pi N^2 \times 10^{-8}}{\Sigma \dfrac{l}{A}} \right) \qquad \text{henries}$$

or $L = \mu_e L_0$, where L is the self-inductance in henries and L_0 the air inductance in henries ($\mu = 1$ for air). For small air gaps we can consider μ_i to be approximately equal to μ_e (the effective permeability of the pot core), and hence L does, indeed, equal $\mu_e \times L_0$.

From the foregoing it can be seen that a manufacturer can specify a particular μ_e for a given type of core material and core dimensions (inclusive of gap).

4.4.3 Pot-Core Designs

A common application for broadband transformers that utilize toroid or pot cores is depicted in Fig. 4-25. For a practical discussion we can consider T1 as a balun transformer, since it converts a single-ended source to a balanced load. This type of product detector offers high dynamic range

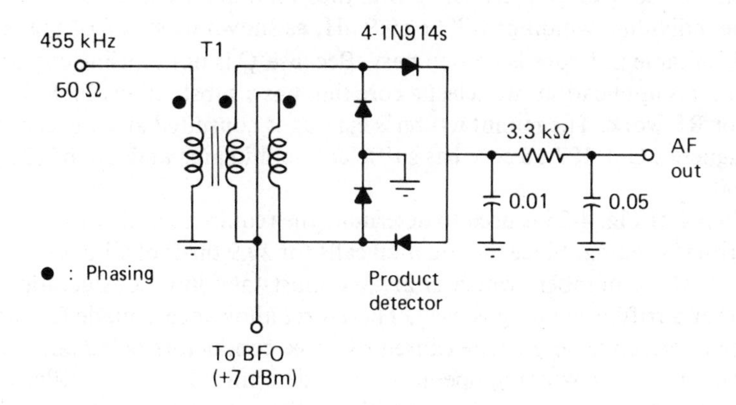

T1: Trifilar wound (broadband)

1. $X_L \geqslant 4 \times 50\ \Omega \therefore X_L \geqslant 200\ \Omega$

2. $L_{min} = 69.9\ \mu H$ ($L_{\mu H} = \dfrac{X_L}{2\pi f}$)

3. Select a suitable pot core.
   Magnetics G-41107-16, 11 x 7 mm
   $\mu_e = 120$, $A_L = 160$. Bobbin area = 0.00785 in.^{2}

4. Determine N
   $N = 1000\sqrt{L_{mH} \div A_L} = 1000\sqrt{0.0699 \div 160} = 20.9$ turns

5. Wind 21 trifilar turns of No. 30 enam.
   wire on bobbin and assemble core.
   (Combine the 3 wires and twist 10
   times per inch before winding bobbin)

Figure 4-25 A typical design procedure is given here for using a pot core transformer in a small-signal application.

which is well beyond the B_{sat} of the pot core. A trifilar winding is placed on the core bobbin after first twisting the three wires approximately 10 times per linear inch. Each wire has a different color of insulation to make identification of the windings easy when connecting them to the circuit.

The four switching diodes in the circuit are small-signal types that should be chosen for similar forward resistances. Hot-carrier diodes are a better choice than ordinary silicon types, but good performance can be had after careful matching of a set of 1N914s or similar types. An RF *RC* filter is used at the detector output to prevent energy other than audio frequency from reaching the subsequent stages of the composite equipment.

As has been the rule throughout this book, the X_L of the windings should be four times or greater the characteristic impedance of the circuit to which they connect. Since we are dealing with a 50-Ω source impedance, X_L should be 200 Ω or greater. Therefore, the minimum acceptable inductance of the individual windings will be 69.9 μH, as shown by step 2 of Fig. 4-25.

A suitable pot core is chosen next. Because Q is not a prime consideration in this application, we select a core that has a substantially high A_L factor for RF work. The manufacturer's catalog is consulted and we learn that a Magnetics G-41107-16 core has sufficient bobbin area and a μ_ε of 120 (A_L = 160).

Step 4 of Fig. 4-25 is used to determine the required number of turns for the trifilar winding. Since the equation calls for 20.9 turns of wire, we use the nearest whole number, which is 21. We must take into consideration the fact that a trifilar winding is used. Therefore, allowance is made for the increase in effective wire gauge caused by three conductors being laid on the bobbin in a single winding operation. It is determined that 21 trifilar turns of No. 30 enameled wire will be suitable for the bobbin area of 0.00785 in.^{2} This is verified by consulting the graph in Fig. 4-23. In fact, some bobbin space should be left over after the winding is placed on the bobbin. The polarity dots on the circuit diagram of Fig. 4-25 must be observed when connecting T1 to the circuit. If not, the phasing will be wrong and the detector will not function correctly.

4.4.4 Tone Encoder with Pot Core

Pot-core inductors find common application in all manner of audio generators. Perhaps the most ordinary type of circuit in which we might employ a high-Q pot-core inductor is the tone generator. A notable example of this application is seen in the standard telephone Touch-Tone pad, which has nearly replaced the old-fashioned dial system. Several pot-core resonators are used to generate the standard Touch-Tone frequencies. A single active device is used as the oscillator. Various inductances are placed in the frequency-determining part of the oscillator circuit by means of the pushbuttons on the Touch-Tone pad.

Other uses for this kind of tone generator are numerous. One example is the "tone-on" or access tone for VHF and UHF repeaters used by some of the land mobile radio services and by radio amateurs. The operator in his or her automobile must key a specified audio-frequency tone when the mobile transmitter is actuated. The tone is decoded at the repeater, thereby directing the repeater transmitter to turn on and be available for service until the duty is performed. When the operator is finished with the repeater, it shuts down and awaits another tone encoding before it is again actuated.

The frequency tolerances established for the tone decoders at repeaters are fairly rigid. Therefore, it is imperative that the tone-encoder oscillators be stable and reliable at all times. Pot-core inductors have proven reliable for many years, and are therefore excellent for the application.

Figure 4-26 contains the circuit of a typical tone encoder device. A device of the type shown is suitable for use with telephone answering ser-

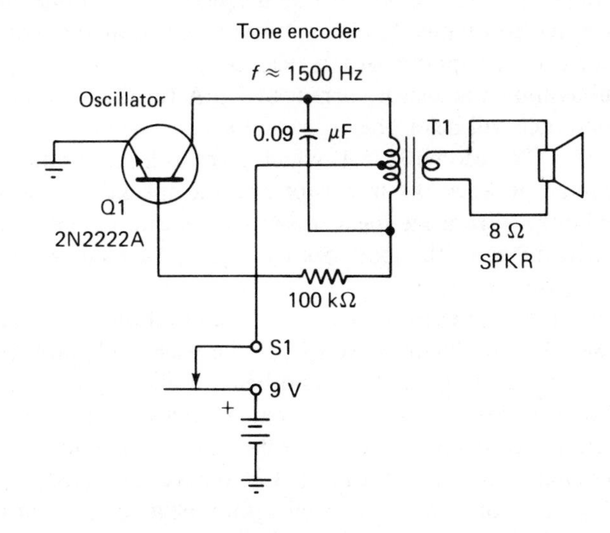

1. T1 pri. = 88 mH. $Z_{pri} \approx 3\ k\Omega \therefore$ Turns ratio = 19:1

2. Select suitable pot core. Magnetics G-42823-X1, 28 x 23 mm. $\mu_e = 307$, $A_L = 1000$. Bobbin area = 0.0910 in^2

3. Determine $N_{pri.}$

$$N = 1000\sqrt{88 \div 1000} = 296 \text{ turns No. 28 enam. wire}$$
tap at 148 turns.

4. Determine N_{sec}. $N = 0.05N_{pri} = 15$ turns No. 28 enam. wire

Figure 4-26 Pot cores are commonly used in encoding oscillators of the type seen here. Design details are given.

vices. The operator can dial or tone up his office or personal phone from some remote location. He or she can then place the speaker of the encoder near the telephone transmitter and command the answering service (unattended) to rewind its tape, then play back any messages that may have been recorded during the operator's absence.

A simple bipolar-transistor tuned base/tuned collector oscillator is used at Q1 of Fig. 4-26. T1 is a pot-core inductor that is wound for an inductance of 88 mH. There is nothing especially sacred about the inductance value: it happens to be a relatively common one in telephone circuits that use toroidal inductors. The circuit is shown with a standard capacitance value of 0.09 μF in parallel with the primary of T1. If 1500 Hz was the precise frequency desired, an adjustable pot core could be used for putting the oscillator exactly on frequency. Alternatively, a high-capacitance trimmer might be used in parallel with the 0.09-μF capacitor.

The oscillator is keyed on and off by means of S1, a momentary switch. Encoding is effected by merely placing the speaker near the mouthpiece of the telephone (or microphone of a transmitter).

We shall assume a collector current of 3 mA for Q1. Therefore, the dc collector resistance will be on the order of 3000 Ω. The impedance transformation ratio for T1 becomes 375:1, which gives us a turns ratio of 19.36:1. A precise match between the transistor and the speaker is not important. The required output from the speaker is modest, negating the need for maximum power transfer to the load. Step 1 of Fig. 4-26 shows that we have chosen a turns ratio of 19:1.

Step 2 of the design exercise requires some intuition. We are aware that to develop 88 mH of inductance we will need a reasonably large pot core. It should have a substantially high μ_e, and hence a high A_L factor. However, the core chosen needs to permit a Q that is high enough to ensure good oscillator action. At this juncture we are also aware that a large area core in a low-power circuit such as that of Fig. 4-26 relieves our need to ponder the B_{max} profile of the core: we will be well within the area of linear operation.

After studying the manufacturers' literature we have selected a Magnetics G-42823-X1 pot core. Equivalent types made by other manufacturers would be entirely suitable also. The core measures 28 × 23 mm, has a μ_e of 307 and an A_L of 1000. The published curves indicate that the Q_u of the inductor should be in excess of 200 at 1500 Hz. The bobbin area for this core is 0.0910 in.^{2}.

Using the standard equation for N we learn that 296 turns of wire are required to develop 88 mH of inductance (step 3 of Fig. 4-26). No. 28 enameled wire will be small enough in diameter to enable us to place 296 turns on the bobbin, with ample space remaining for the secondary winding of T1.

The design is concluded by moving to step 4. A layer of thin insulating tape is place over the completed primary winding of T1; then the secondary is laid on the bobbin. We shall need 15.57 turns of wire for the secondary in order to realize the 19:1 turns ratio of step 1. A 20-turn winding is employed.

4.4.5 Pot Cores in Filters

An example of a five-pole low-pass audio filter is offered in Fig. 4-27. The same core number as that specified for the circuit of Fig. 42-6 will be used to obtain the inductances of L1, L2, and L3, which are 17 mH (L1 and L3) and 24 mH (L2).

In step 2 of Fig. 4-27 we learn that L1 and L3 will consist of 130 turns each, wound on bobbins for Magnetics G-42823-X1 pot cores. L2, according to the equation in step 3, contains 155 turns of wire.

Finally, the wire size is chosen in accordance with the 0.091-in.[2] bobbin area. No. 24 enameled wire will suffice. The self-shielding properties of the

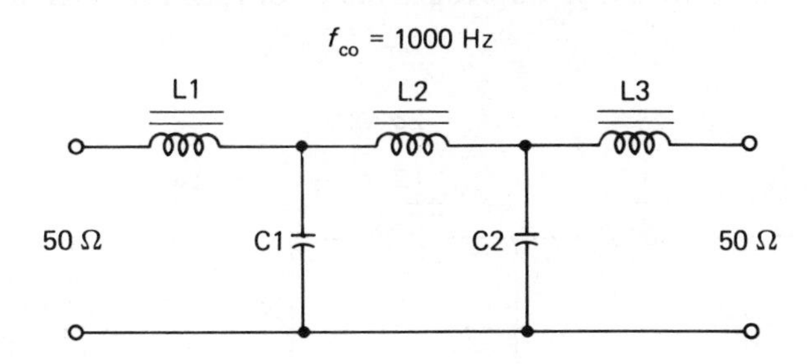

5-Pole low-pass audio filter

$X_{L1,\,L3} = 106.8\ \Omega = 17$ mH $\qquad X_{C1,\,C2} = 45.9 = 3.47\ \mu$F

$X_{L2} = 150.34\ \Omega = 24$ mH

1. Select high-μ pot core
   Magnetics G-42823-X1, $\mu_e = 307$,
   $A_L = 1000$, 28 x 23 mm. Bobbin in area = 0.091 in.^{2}

2. Calculate turns for L1, L3

   $$N = 1000\sqrt{L_{mH} \div A_L} = 1000\sqrt{17 \div 1000} = 130 \text{ turns}$$

3. Calculate turns for L2

   $$N = 1000\sqrt{24 \div 1000} = 155 \text{ turns}$$

4. Determine wire size vs. bobbin area, per Fig. 4-23.
   Use 130 turns No. 24 enam. wire for L1, L3 and
   155 turns No. 24 enam. for L2

Figure 4-27 Many filters contain pot-core inductors. The five-pole low-pass filter seen here is one example. A basic design procedure is included.

pot cores will prevent the need for using partitions between the filter sections.

A seven-pole low-pass filter is illustrated schematically in Fig. 4-28. It has an f_{co} (cutoff frequency) of 2.7 MHz. Therefore, the core material for the three inductors must be selected accordingly. The improper core would degrade the Q and render the filter ineffective.

Step 1 of Fig. 4-28 specifies a Ferroxcube 1107PA25 pot core of 4C4 material. The dimensions are given in inches and millimeters (11 × 7 mm). The μ_e is 19 and the A_L is 25. The manufacturer's Q curves indicate than the core material is suitable for use at 2.7 MHz and lower.

N for L1 and L3 is found in step 2, again using the standard equation for N versus A_L. 12.6 turns are specified for the end coils of the filter.

The calculations for L2 are conducted in step 3 of Fig. 4-28. The resultant number is 13.8. A wire gauge of No. 22 will be used for all three inductors. The larger wire gauge will reduce the R of each coil, thereby enhancing the Q.

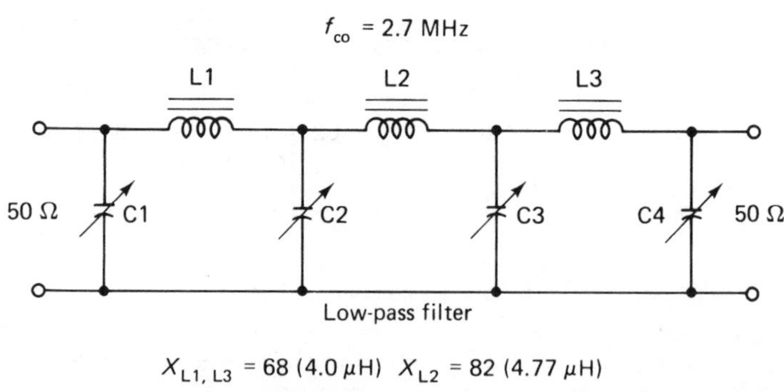

$X_{L1, L3} = 68$ (4.0 μH) $X_{L2} = 82$ (4.77 μH)

$X_{C1, C4} = 69$ (854 pF) $X_{C2, C3} = 30$ (1966 pF)

1. Case: Ferroxcube 1107PA25-4C4

   $\frac{7}{16}$ in. (11 mm) OD x $\frac{9}{32}$ in. (7 mm) high

   $\mu_e = 19$, $A_L = 25$. $Q_\mu \approx 150$ at f_{cO}

2. $N_{L1, L3} = 1000\sqrt{L_{mH} \div A_L}$

   $= 1000\sqrt{0.004 \div 25} = 12.6$ turns

3. $N_{L2} = 1000\sqrt{0.00477 \div 25} = 13.8$ turns

4. Select wire gauge
   (No. 22 enam. will fit bobbin
   for turns specified)

Figure 4-28 Pot-core inductors are suitable for tuned filters of the variety depicted here. Since the f_{co} of this filter is 2.7 MHz, the proper kind of core material must be chosen to ensure proper Q (see the text).

5

PERMANENT-MAGNET DATA

Chapters 1 through 4 dealt specifically with magnetic-core materials that were not magnetized permanently. This chapter addresses another part of the core-material family—those cores which are permanently magnetized for a host of modern applications. Among the common uses of magnets are loudspeakers, phonograph cartridges, magnetic latches, and meters. Other uses include traveling-wave tubes, magnetrons, klystrons, generators, and alternators. This list is by no means complete, but it does illustrate the widespread use of magnets in the electrical and electronics industry.

In this section the author has borrowed heavily from a booklet entitled, *Permanent Magnet Guidelines,* produced by MMPA (Magnetic Materials Producers Association), with their kind permission. The pamphlet was so well written, and the contents of such importance, that it was selected as a valuable reference in this book.

Another book by MMPA is available on the subject matter of this chapter. It is *Testing and Measurement of Permanent Magnets.* It is recommended highly as a supplemental text for students, technicians, and engineers.

5.1 The Nature of Permanent-Magnet Materials

French physicist Pierre Weiss, some 50 years ago, supplied the central thought that explained the observed properties of ferromagnetic materials and today still provides the basis of our highly sophisticated body of knowledge that explains quite satisfactorily the observed properties and provides an intelligent guide for the search for improved materials. Weiss postulated that a ferromagnetic body must be composed of small regions or domains each of which are magnetized to saturation level, but the direction of the magnetization from domain to domain need not be parallel. Thus, a magnet when demagnetized, was only demagnetized from the viewpoint of an observer outside the material. Man-made fields only serve as a control in changing the balance of field energy within a magnet.

The inherent atomic magnetic moment associated with such elements as iron, cobalt, and nickel is believed to originate or result from a net unbalance of electron spins in certain electron shells. For example, in iron in the 3*d* shell there are more electrons spinning in one direction than in the other. Having an inherent atomic magnetic moment is a necessary but not a sufficient condition for ferromagnetism to be exhibited. Additionally, there must be cooperative interatomic exchange forces that maintain neighboring atoms parallel. Little is known of the exact nature or magnitude of these forces, but observations suggest they are electrostatic, and it has been pointed out that in ferromagnetic materials the ratio of interatomic distance to the diameter of the shell in which the unbalance exists is unusually large compared to this ratio in materials which do not exhibit ferromagnetism. Also note that these exchange forces produce magnetostrictive effects and are associated with the crystallographic structure of magnetic materials in such a way as to produce a directional dependence of magnetization with respect to the crystal axis. Figure 5-1 illustrates this relationship in iron; the easy axis of magnetization is the cube edge of (100) direction. In Figure 5-2 an exploded view of a ferromagnetic volume shows the order of size for the various regions.

We can view the magnetic domain as a region in which the atomic moments cooperate to allow a common magnetic moment which may be rotated in man-made external fields. Domain size, not a fundamental constant of physics, varies widely depending on: composition, purity, and state of strain of the material; and on some very important energy relationships. Figure 5-3 shows a boundary region between two domains. This boundary region and its significance were first proposed by Bloch. The Bloch wall is a transition region containing many atomic planes. The 180° change in magnetization will want to occur over a considerable distance to minimize the potential energy in the wall. However, the width of the wall will be restricted because of the restraining influence of crystal anisotropy (directional dependence of magnetism with respect to crystal axis). Figure 5-4 illustrates an additional energy relationship which influences the size of the domain and involves the magnetostatic or field energy surrounding a magnetized volume. A magnetized volume tends to subdivide. It will be energetically possible for subdivision to occur as shown in Figure 5-4 until the decrease in magnetostatic energy is less than the potential energy associated with the Bloch wall foundation. At this point we might say that

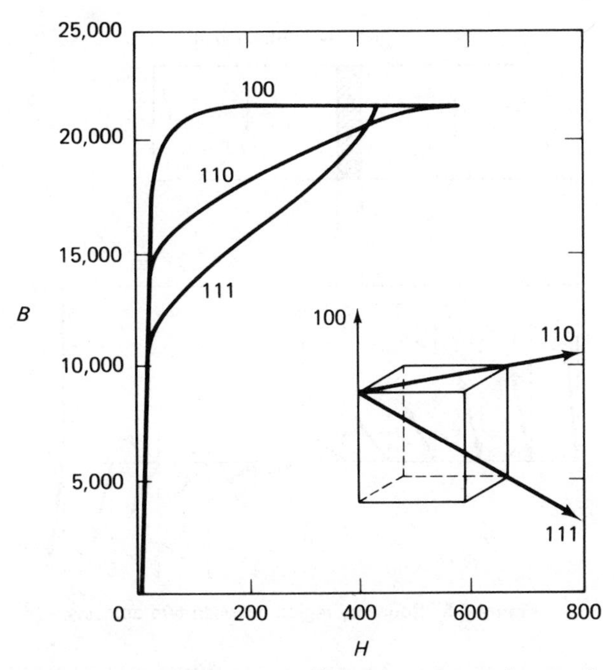

Figure 5-1 Exchange forces affect the structure of magnetic materials in such a way as to produce a directional dependence of magnetization with respect to the crystal axis. In iron (shown) the cube edge of (100) direction is the easy axis of magnetization.

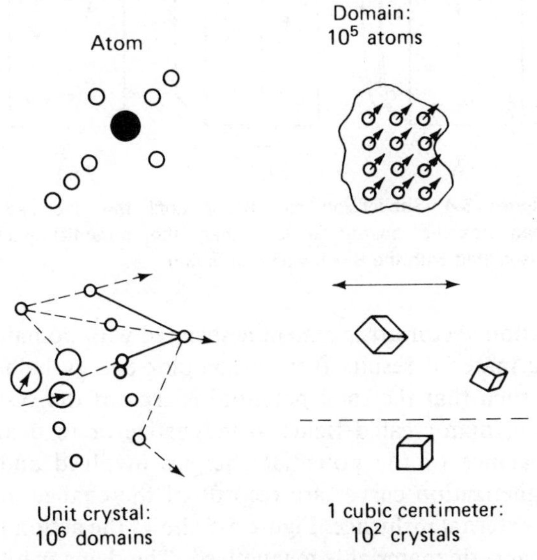

Figure 5-2 Exploded view of a ferromagnetic volume.

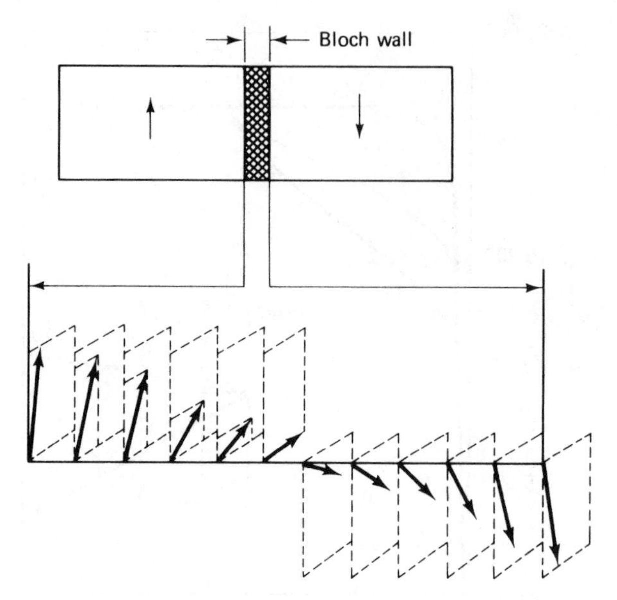

Figure 5-3 Boundary region between two domains.

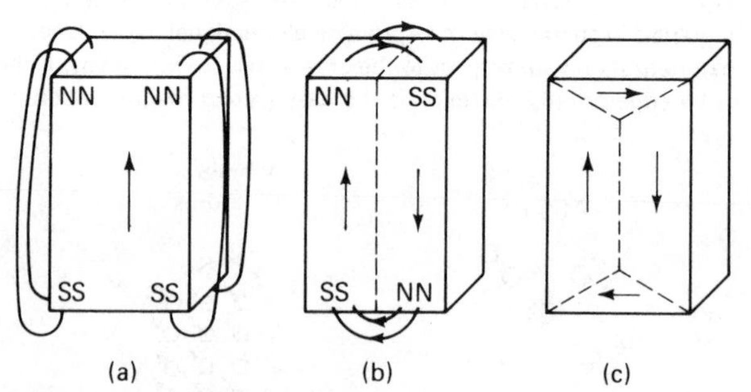

Figure 5-4 Subdivision can occur until the decrease in magnetostatic energy is less than the potential energy associated with the Bloch wall foundation.

the magnetization vector arrangement associated with domain volumes in a ferromagnetic material results from a complex energy balance. Their arrangement is such that the total potential energy of the system is a minimum. External, man-created fields to magnetize or to demagnetize only disturb the balance of the potential energies involved and our familiar S-shaped magnetization curves are records of the change in balance with respect to the external influence. Figure 5-5 shows the action pictorially as a bar of ferromagnetic material is magnetized. The demagnetized condition, (A), results from an internal arrangement with mutually canceling direc-

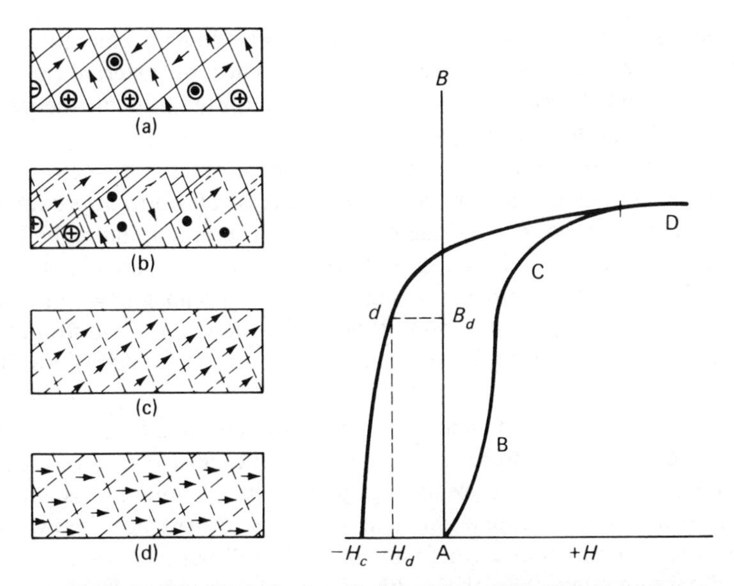

Figure 5-5 Magnetization of a bar of ferromagnetic material.

tions of magnetization vectors. In region (B), with low values of external field, the action is primarily one of domain boundary stretching, usually around imperfections. This is a reversible process [a reversible magnetization process is one in which the magnetization vectors reorient to their original position after the field (H) is removed]. As the field is increased, region (C) domain boundaries break away and move through the material. The more favorably oriented regions grow at the expense of their less favorably oriented neighbors. A large increase in magnetic induction occurs, an irreversible process [in which the magnetization vectors tend to keep their new position after (H) is removed]. In region (D) at still higher values of magnetizing force, the magnetization vectors are rotated against the forces of strain and crystal anisotropy into alignment with the direction of the applied field, and saturation occurs. Removing the magnetizing force causes some relaxation; the domains rotate back to the easy direction of magnetization (a reversible process). This relaxation can be minimized by making the direction of easy magnetization coincident with the direction of magnetization desired.

Subjecting the magnet to a demagnetizing force returns the domain boundaries to a condition similar to their original positions in (A) and hence the magnet is demagnetized. In normal use a permanent magnet operates in the second quadrant of the hysteresis loop and since the magnetization force is negative the usable induction will be $B = B_1 + (-H)$ and the magnet will operate at some point such as Figure 5-5d, where a magnetic potential $-H_d$ per unit length and induction $+ B_d$ per unit section will be established. For outstanding permanent magnets we want to make domain wall motion and rotation of the magnetization vectors difficult. The more external energy required to orient the system, the more will be required to demagnetize and hence a better permanent magnet.

Figure 5-5 describes the magnetization process satisfactorily for magnetic materials having coercive force value of H_0 up to approximately 300 Oe. The coercive force of the early carbon, tungsten, and cobalt–steel permanent magnets is believed to be a result of impeding domain wall motion. These quench-hardened magnets have nonmagnetic inclusions building up at the domain boundaries, obstructing wall motion. This mechanism is believed to be of significance only at relatively low field strength.

Today's modern permanent magnets exhibit coercive forces well above the level explained by domain wall motion. One must therefore look for magnetization mechanisms requiring greater energy input. A significant milestone in understanding permanent magnet properties occurred with the suggestion of Frenkel and Dorfman that, if small particles were prepared with dimensions less than the width of domain boundary, such particles would contain no boundaries. This explanation forms the central concept in fine-particle magnet theory and provides us with a satisfactory picture and explanation of high coercive force permanent magnets such as Alnico, ferrite, and fine-particle iron–cobalt types. The dimensions of a single domain volume are predictable from a consideration of wall energy and magnetostatic energy. For a sphere the wall energy is proportional to the cross section or to the square of the radius. The magnetostatic energy is proportional to the volume or to the radius cubed. A critical radius exists where these two energy values are equal, for this value and below, it is energetically impossible for a domain boundary to exist. Without domain boundaries the magnetization of a permanent magnet can be changed only by rotation of the magnetic moments associated with each domain volume. This process requires higher energy input than domain boundary motion. The degree of difficulty in reversing the magnetic moments or the coercive force of a single domain system depends on the forces of anisotropy, or the forces which give a directional dependence to a domain's magnetic moment. Crystal and strain anisotropy have been mentioned previously; additionally, shape anisotropy is of great importance. In fine-particle magnets, shape anisotropy arises from the fact that, in an elongated single domain, the magnetization along the major axis is the easy axis and involves minimum energy. To rotate the magnetization vector and magnetization to the minor axis involves additional energy. Hence, elongated particles exhibit higher coercive force than spherical particles. In service, the permanent magnet is a unique component in the energy-conversion cycle. When the permanent magnet is magnetized, energy is dissipated in changing the force balance of magnetization vectors. For the permanent magnet to establish external field energy, input is required to change the vector orientation near the poles and establish the permanent magnet's magnetomotive force. Energy is only involved in changing a magnetic field, not in maintaining one. Hence, the field energy established via the permanent magnet is independent of time unless the permanent magnet is subjected to additional energy input (heat energy or demagnetizing field energy, for example). A permanent magnet in a stabilized condition is a reversible medium of energy transformation. Potential energy of the permanent magnet is composed of both internal and external field energy. Permanent magnets often operate over a dynamic cycle in which energy is converted from an electrical or mechanical form to the magnetic field and then returned to the original form.

5.2 Elementary Permanent-Magnet Relationships

The basic problem of permanent magnet design is to estimate the distribution of magnetic flux associated with a specific magnetic circuit geometry that may include permanent magnets, air-gaps, and soft-steel flux conduction elements. This is a very difficult problem and only approximate solutions are feasible based on certain simplifying assumptions. Determining how suitable or appropriate these assumptions are in a particular case requires appreciable experience. This section attempts to present only a few basic concepts and inter-relationships. It should be emphasized that obtaining an optimum magnet design in a particular situation involves experience and trade-offs between many different information inputs involving both the permanent magnet and the equipment in which it is used.

5.2.1 The B-H Curve

The basis of design is the B–H curve, or hysteresis loop, typical of all magnetic materials, which is obtained from measurement on a specimen under closed-circuit conditions (Figure 5-6). Solid lines represent the normal form of the curve, and the arrows indicate the direction in which the curve traverses a symmetrical cycle of magnetization. The dashed curve shows the intrinsic hysteresis loop. Although in the usual utilization of permanent magnets the normal curve is used for analysis, it is desirable to note the difference between these two curves, because it is very significant in permanent magnets. These two curves are related by the equation $B = B_i + H$, where B is the normal induction, B_i the intrinsic induction, and H the field strength. In the second quadrant of the hysteresis loop H is negative, and hence B_i is greater than B. In a permanent magnet the field strength, H, is directly opposed to the intrinsic induction B_i and B is the net induction available for external use. It should be noted that this is in contrast to the electromagnet, in which B_i and H are in the same direction.

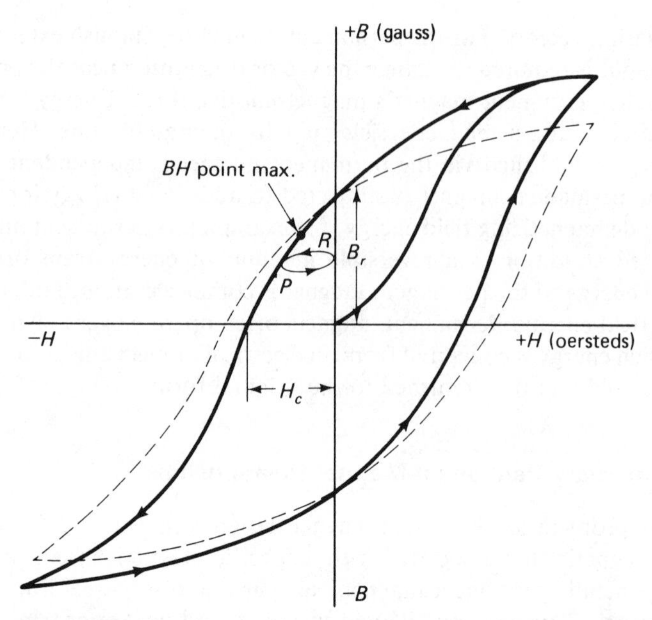

Figure 5-6 The B–H curve, or hysteresis loop is obtained from measurement on a specimen under closed circuit conditions. Solid lines indicate the normal form of the curve; arrows indicate the direction in which the curve traverses a symmetrical cycle of magnetization. Dotted curve shows intrinsic hysteresis loop.

The most important characteristics of permanent magnetic materials, residual induction, B_r, coercive force, H_c, and maximum energy product, $(B_dH_d)_m$, may be obtained from the demagnetization quadrant of the hysteresis loop. The residual induction, B_r, is defined as the magnetic flux density of the material in a closed circuit when the applied magnetizing field, H_s, has been removed, while the coercive force, H_c, is the demagnetizing force, which reduces the retained flux to zero. $(B_dH_d)_m$, a criterion of magnetic properties, may be obtained by constructing the energy product curve, that is, the product of abscissa H and ordinate B plotted against the value of (B) or (H) at appropriate points on the demagnetization curve. A smooth curve drawn through the plotted points will commence at zero, the origin, rise to a maximum point, $(B_dH_d)_m$, and fall again to zero.

5.2.2 Recoil Loops

If, at any point such as P in Figure 5-6, the cyclic change in H is reversed in direction, the flux density changes along a curve such as PR. If, from R, the field strength H retraces its values, the induction traverses the upper half of the loop (PR) until it again reaches P, after which the major hysteresis loop is followed. The loop (PR) may be coincident with the major loop. Such a loop is known as a *recoil loop,* the average slope of which is fairly constant whatever the point of origin. This slope is known as the recoil (or reversible) permeability (μ_{re}). It is sufficiently accurate for engineering purposes to regard the loop as a straight line.

5.2.3 *Working Point of a Magnet (Static Applications)*

A permanent-magnet system usually has to provide magnetic flux in an air gap. The presence of this gap in the circuit produces a magnetic field opposing the flux in the magnet material. In other words, magnetic energy exists in the gap, and in providing this energy the magnet is correspondingly self-demagnetized. Therefore, if the induction B of the magnet is positive, the field strength H inside the magnet must be negative, which means that the magnet works in the second or fourth quadrant of the hysteresis loop (i.e., the demagnetization curve). If the magnet is demagnetized by the gap alone, the operating point is on the demagnetization curve. However, if after saturation, the magnet undergoes additional demagnetization (e.g., by a "stabilizing" field) the operating point is inside the demagnetization curve, on a recoil loop.

The whole volume of a magnet does not operate at the same point but, for simplicity, it is usual to assume that it does. The error inherent in this assumption is greatest when the air gap is long, but much less with a short air gap or when the magnet has substantial soft-iron pole pieces. For a magnet of cross section A_m (cm^2) and length l_m (cm), the total flux provided in external space is B_dA_m and the total magnetomotive force across the ends of the magnet is H_dl_m. Most design problems can be reduced to a specification of the total flux and the total magnetomotive force to be provided by the magnet. For a given alloy, the total volume of magnet material required (A_ml_m) is inversely proportional to (B_dH_d) at the chosen operating point, which enables the length and section of the magnet to be fixed. For economy, it is usual to choose the operating point, where the energy product has its maximum value for the material but, if the magnet dimensions thus obtained are impractical, another operating point or a different material must be selected.

5.2.4 *The Magnetic Circuit*

In designing a magnetic circuit, it is necessary, in addition to the normal assessment of the total flux and total magnetomotive force required, to assume that the magnetic circuit can be divided into various elements and treated like an electric circuit by applying the magnetic analog of Ohm's law:

$$\text{flux} = \frac{\text{magnetomotive force}}{\text{reluctance}} \tag{5.1}$$

The main difficulty in applying this formula is to assess the reluctance. For short gaps of large area the reluctance is

$$\frac{\text{gap length (cm)}}{\text{gap area (cm}^2\text{)}} \tag{5.2}$$

In practice, the effective reluctance is less than this value, since part of the flux is leakage. This is equivalent to a proportionate increase in the gap area, which may be considered as a leakage factor *(F)*. This factor is usually between 1.5 and 20, and, although it may be calculated approximately by taking the reciprocal of the sum of reciprocals of all leakage path reluctances, it is more reliably based on previous experience of comparable designs.

Similarly, it is frequently advisable to apply a correction factor to the gap length to allow for the losses introduced by curvature of flux path, subsidiary joints in the magnetic circuit, and loss in pole-piece materials. This factor (*f*) is usually much smaller than the area factor, and is normally between 1.1 and 1.5. For a gap of section A_g (cm²) and length l_g (cm), the corrected reluctance is

$$\frac{l_g f}{A_g f} \tag{5.3}$$

The full circuit equation is

$$B_d \cdot A_m = \frac{H_d l_m A_g f}{l_g f} \tag{5.4}$$

where B_d and H_d are the values at the assumed point of the magnet of section A_m and length l_m. This can be transposed as

$$\frac{B_d}{H_d} = \frac{l_m A_g f}{A_m l_g f} \tag{5.5}$$

The reciprocal of reluctance ($A_g F/l_g f$) is known as the *permeance.*

The whole right-hand side of equation (5.5) is known as the unit permeance and may be represented graphically by a straight line passing through the origin of a B–H curve. Several such lines are shown on Figure 5-7. Every design has a corresponding operating line, and its slope depends on the dimensions of the magnet and the gap, with correction of leakage factors. The slope is independent of the alloy of which the magnet is made. The intersection of this line with the demagnetization curve of the alloy gives the working point of the magnet.

The foregoing gives the ratio l_m/A_m for the magnet and it remains to fix either l_m or A_m. If the total flux required from the magnet (including leakage) is known, then A_m can be determined, as this total flux is $B_d \cdot A_m$. If the field strength H_g in the specified air gap is known, then

$$H_d l_m = H_g l_g f \tag{5.6}$$

Equations (5.5) and (5.6) together give the section and length of the magnet required in an alloy of known demagnetization curve and of chosen values of B_d and H_d. These values may be: at the $(B_d H_d)_m$ point, to give minimum volume of material; at some other point on the demagnetization curve, if practical considerations of size make it necessary; or on a recoil loop, if the conditions of operation make this desirable or inevitable. If the magnet is magnetized before being attached to the pole pieces, it may work on a recoil loop. In a motor or generator the magnet has no fixed working point, but traverses a recoil loop.

In many uses of the permanent magnet, a well-defined magnetic circuit is not involved (e.g., a simple bar magnet). Under such conditions the operating point on the demagnetization curve is determined primarily by the dimension ratio l_m/dia of the bar.

5.2.5 The Permanent Magnet in an External Field

In using a permanent magnet the influence of external fields in changing the flux level is an important consideration. The geometry of the magnet assembly sets the slope of the operating line and has considerable effect on the interaction of the permanent magnet and the external field. To predict

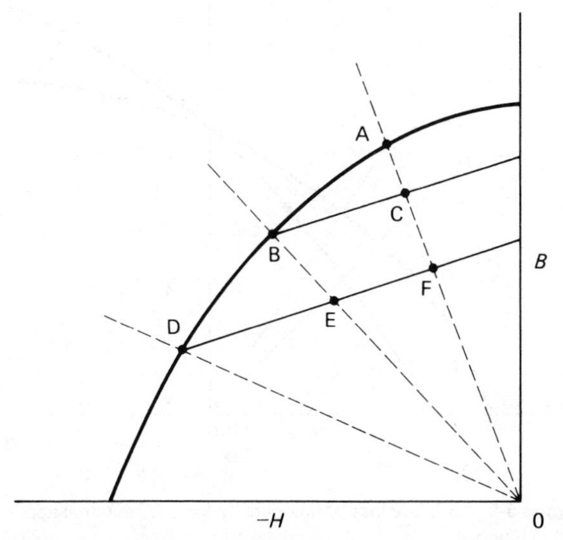

Figure 5-7 Typical demagnetization curves showing various operating slopes.

the change in flux density of a permanent magnet in the presence of an external field, we must have both the normal and intrinsic curves of the permanent magnet material. If the operating line slope or unit permeance B_d/H_d is known, a line with a slope of $B_d/H_d + 1$ is drawn through the origin, (Figure 5-8). The magnitude of an external field influence $-H_a$ is laid off parallel to $B_d/H_d + 1$ and its intersection with the intrinsic curve is projected down to the normal curve to yield the new level of flux density in the presence of this external influence. In Figure 5-8, ΔB is the loss of flux density for $-H_a$ external applied field influence.

5.3 Magnetization

A permanent magnet, to exhibit its full properties, must be magnetized to saturation. Partial magnetization results in reduced properties, and the efficiency and stability of the permanent magnet are consequently impaired.

The permanent-magnet manufacturer generally ships demagnetized magnets to the user. The main reason is that the permanent magnet is usually incorporated into a magnetic circuit and the operating flux levels can only be obtained by magnetization after assembly into the magnetic circuit. Additionally, shipping costs of magnetized permanent magnets are greater and the danger of contamination due to pickup of magnetic particles is lessened by shipping and handling demagnetized magnets.

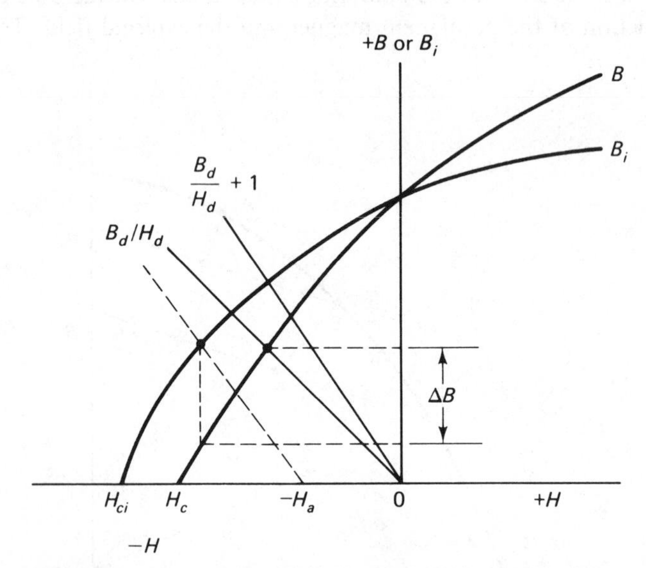

Figure 5-8 B is the loss of flux density for $-H_a$ external applied field influence.

5.3.1 Conditions for Complete Saturation

To fully saturate a permanent magnet requires careful consideration of the magnitude, the shape, and the time duration of the applied field. In a previous section, the term "self-demagnetization" was introduced. In magnetization the free pole or self-demagnetization concept is extremely important because the self-inflicted, self-demagnetization is a maximum for a permanent magnet during magnetization.

The saturation magnetizing force (H_s) of a permanent magnet material may be taken as four to five times its intrinsic coercive force. In designing a magnetizer it must be remembered that the value of (H_s) must be the net effective value acting on the magnet, and the presence of self-demagnetization or shuntings effects will necessitate additional magnetizing field strength to compensate for these factors. The self-demagnetization influence will be a maximum for the case of a bar magnet having a small ratio of length to cross-sectional area magnetized in a solenoid. For example, a relationship between net effective working force as seen by the permanent magnet (H_s) and the field strength of the solenoid ($H_{sp;}$) may be expressed as follows:

$$H_s = H_{sol} - K(B_s - H_s)$$

where K is a demagnetization constant depending on magnet geometry (Table 5-1); B_s is the flux density associated with field strength H_s. An example, perhaps, best illustrates the importance of free poles and the self-demagnetization in magnetizing problems. Consider two Alnico 5 rods, one having a l_m/D ratio of 0.1 and the second a l_m/D ratio of 10.0. To saturate Alnico 5, $H_s = 3000$ and $B_s = 17{,}000$ G. Using the relationships described, the rod with a $l_m/D = 0.1$ would require a value of $H_{sol} = 14{,}900$ Oe, the rod with a $l_m/D = 10.0$ would require a value of $H_{sol} = 3240$ Oe. This example shows that the free-pole influence of the magnet geometry having heavy self-demagnetization can be minimized by the use of soft-iron parts to complete the magnetic circuit.

TABLE 5-1 K for rods and bars.

Dimensional ratio (length/diameter[a])	0.0	0.1	0.2	0.5	1.0	2.0	5.0	10.0
K	1.000	0.850	0.730	0.472	0.270	0.140	0.040	0.017

[a] Diameter of bar may be taken as $2\sqrt{\text{area}/\pi}$.

In addition to having an adequate magnitude of magnetizing force, it is necessary that the shape of the applied field coincide with that of the magnet as closely as possible. Modern permanent magnets are low in permeability and their presence in a field will not, to any great extent, shape the field. Current-carrying conductors combined with soft-steel members can be arranged to give almost any field contour desired. As a general statement, it is not objectionable to use magnetizing force in excess of the minimum required. However, when the field contour is very dissimilar to that of the magnet, magnetization at some angle from the desired axis may result. The net effect would be one of partial or incomplete magnetization.

The duration of the magnetizing field is often a factor. We know that the rise of magnetic induction is essentially instantaneous in thin magnetic films, but in a metallic permanent magnet of appreciable cross section, induced eddy currents can influence the rise of magnetization. In equipment such as electromagnets, the rise of magnetizing current is not instantaneous, owing to inductance. Many large electromagnets require buildup times in the order of seconds.

5.3.2 Magnetizing Equipment

Steady-state fields produced by electromagnetic yokes and permanent-magnet structures are convenient forms of equipment for magnetizing simple-shaped rods, bars, and rings. The conventional electromagnet magnetizer has one pole movable with respect to the other, is usually operated from a full-wave rectifier, and is designed for intermittent operation. Figure 5-9 shows a typical arrangement.

Principal design considerations for an electromagnet to magnetize a particular magnet are:

1. Core section must be adequate to carry the saturation flux.
2. Ampere-turns must be sufficient to produce the total mmf required by the magnet length, the magnetizer yoke, and any self-demagnetization influences present.

Because of the leakage and losses, the core of the electromagnet should be at least two times the area of the magnet to ensure adequate flux level in the permanent magnet. The ampere-turn requirements are typically twice the value computed to saturate the length of permanent-magnet material.

Very small permanent magnets are often magnetized by means of a large permanent magnet. The amount of physical effort and the uniformity problem created by moving the magnetized magnet through a nonlinear field adjacent to the main air gap greatly limit this approach. Since the time interval required is extremely short, magnetizing can be achieved by a current impulse, provided that its magnitude is sufficient to deliver the peak magnetizing force required.

The development of material with a high coercive force and high available energy has led to relatively intricate magnet configurations. The length of magnet limbs has decreased and circular shapes and parallel cir-

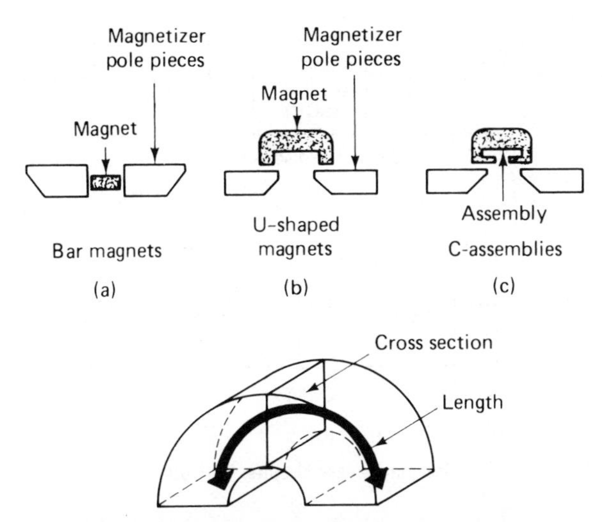

Figure 5-9 When selecting an electromagnetic magnetizer for your application, consider two factors: area of magnetizer core should be about twice cross-sectional area of magnet; for Alnico magnets, ampere turns should equal 6,000 per inch of magnetic length of magnet. A slight surplus of ampere turns is desirable when possible. For ceramic magnets, approximately 20,000 ampere turns per inch are required. (a) Bar magnets; (b) U-shaped magnets; (c) assemblies.

cuits are common. Many of the newer-shaped permanent magnet arrangements cannot be magnetized by placing them in contact with conventional electromagnets. Instead, they are magnetized by the field around a conductor carrying a high current impulse. Some shapes can only accommodate a single conductor threading through the window of the magnetic circuit. Others must be wound with several turns of heavy wire. In fact, many magnet designs are materially influenced by how the magnetizing conductor or conductors can be arranged.

Impulse magnetization has become more popular not only for these newer shapes, but also for nearly all types of permanent magnets, because of the flexibility, low initial equipment cost, and the ability to deliver higher instantaneous currents from conventional ac power mains. Figure 5-10 shows typical impulse magnetizer circuits.

The current required for impulse magnetization around a single conductor can be estimated from $I = 5\ H_s(r)$, where I is the peak current in amperes, H_s the saturation magnetizing force in oersteds, and r the maximum radius of the magnet configuration in centimeters.

Figure 5-11 shows typical conductor arrangements for magnetizing various permanent-magnet configurations. In impulse magnetization, induced eddy currents in metallic magnets can impede flux penetration in

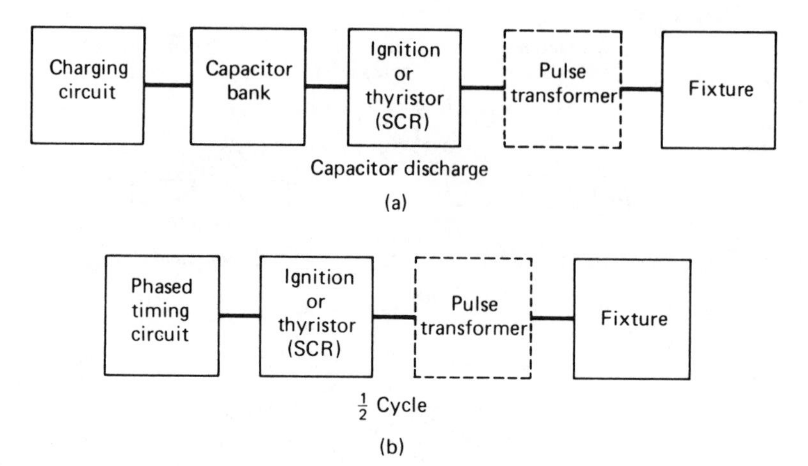

Figure 5-10 Typical impulse-magnetizer circuits. (a) Capacitor discharge; (b) half-cycle.

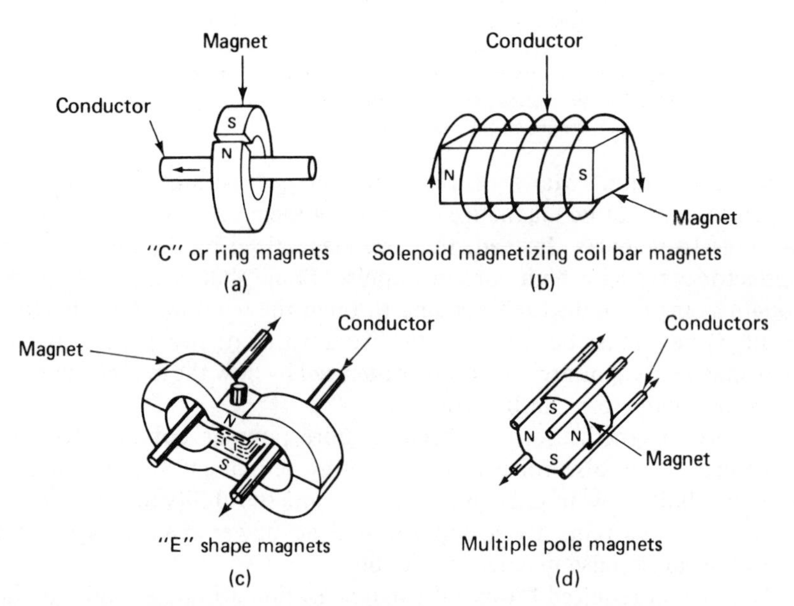

Figure 5-11 Typical conductor arrangements for magnetizing various permanent-magnet configurations. (a) C or ring magnets; (b) solenoid magnetizing coil bar magnet; (c) E-shaped magnets; (d) multiple pole magnets; (e) multiple poles on one face of disk or ring magnet; (f) multiple poles on ID of stator magnet or assembly; (g) two-pole on ID of stator magnet or assembly.

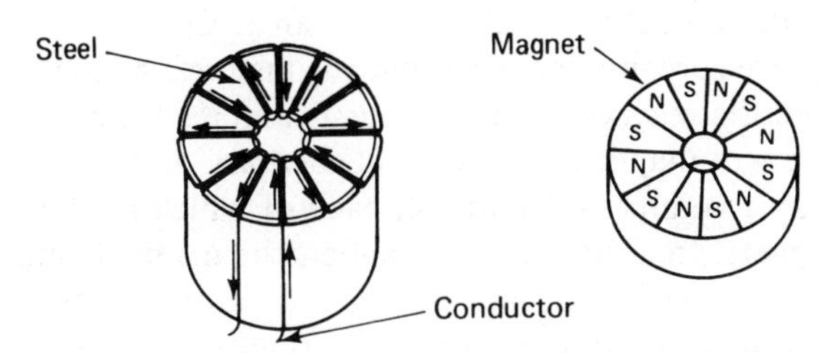

Multiple poles on one face or disc or ring magnet

(e)

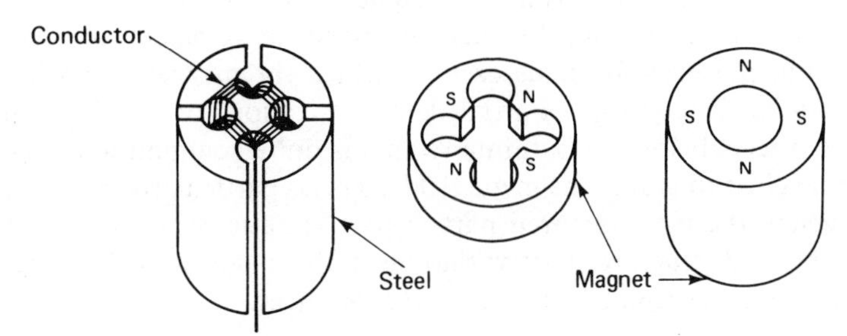

Multiple poles ID of stator magnet or assembly

(f)

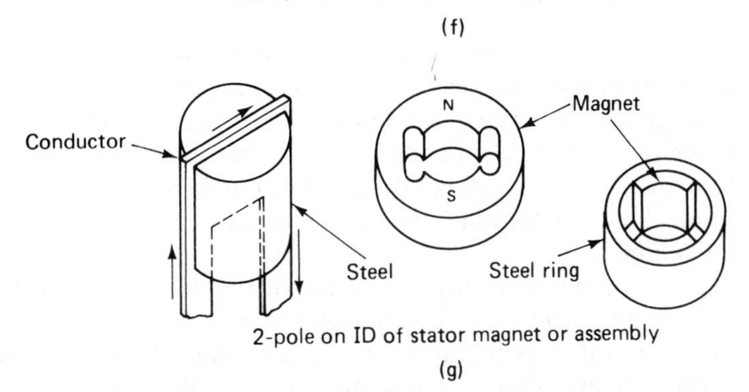

2-pole on ID of stator magnet or assembly

(g)

magnets of appreciable cross section. It is often necessary to use several current pulses to achieve saturation.

How does one know when saturation has been reached? This is a difficult question. The common practice is to use a surplus of magnetizing force, but often one encounters what is believed to be a marginal saturation. Under such conditions it is recommended that the user consult his or her magnet supplier.

5.4 Demagnetization

Demagnetization is defined as reducing to an acceptable level the flux density, in the space surrounding a permanent magnet, at some predetermined distance from the magnet surface. At temperatures below the Curie temperature, permanent magnets consist of elementary domains—microscopic or submicroscopic volumes of material, each of which are tiny saturated permanent magnets. In a magnetized magnet, the magnetic moments of these domains are aligned essentially parallel. In a "perfectly" demagnetized magnet, they are aligned in a statistically random manner. Thus, even in a "perfectly" demagnetized magnet, some external field will be present, if only on a submicroscopic scale. As a generality, demagnetization is handled by the manufacturer of permanent magnets. This is because magnets must be magnetized for testing, then demagnetized for shipment, in most cases. There will, however, be some cases in which the magnet user will find it necessary to demagnetize. Partial demagnetization, or stabilization, to a measured level below the saturated value is quite common in applications requiring close tolerance magnetic fields. This is particularly true in applications where the manufactured part-to-part variations in both the magnet and associated structure cause variations in the magnetic field strength that are outside the acceptable end-product tolerances.

5.4.1 Demagnetization Techniques

AC Fields. The most common method of demagnetizing is to subject the magnet to an ac field of an intensity sufficient to approximately saturate the magnet, followed by gradual reduction of the field to zero, either by withdrawing the magnet to a remote location with the ac power on, or by gradually reducing the ac voltage. In general, the higher the field and the more gradual its reduction, the more nearly you approach ideal demagnetization.

Design of conventional solenoids to provide 60-Hz fields of sufficient intensity to demagnetize modern permanent-magnet materials offers some difficulty if the coil must be run on a semicontinuous basis. Fortunately, only intermittent operation is normally required. Alternatively, one can design the coil, which consists of a few turns of water-cooled conductors, and run it at low voltage and high current, by means of a step-down transformer. This type can be used for continuous duty.

For applications requiring less critical demagnetization, the coil can be connected to a charged capacitor bank. By proper choice of capacitance and coil inductance, the resonant frequency can be made sufficiently low to accomplish demagnetization as a result of the damped oscillatory nature of the current in the coil–capacitor combination. It is difficult to completely demagnetize large metallic magnets by means of ac methods because the eddy currents generated near the surface tend to shield the interior from the ac field. Very high values of the ac field will aid in this regard, as will lower frequencies. AC fields are also applicable to calibration.

DC Fields. Using dc, two methods can be employed to perform demagnetization (although less than ideally). In the first, a magnetizer with a reversing switch and a means to reduce the dc current to zero are required. The magnet is placed in the magnetizer and the field is applied in alternating directions by means of the reversing switch and, at the same time, its magnitude is gradually reduced by, for instance, a rheostat in the dc line or a variable autotransformer in the ac line of the rectifier system. The quality of demagnetization is dependent upon the number of reversals and the field reduction per step.

In the second method, the saturated magnet is exposed to a dc field in a direction opposite to saturation. The magnitude of the field, to be found by experiment, must be such that after exposure, the magnet is demagnetized. This is the least desirable procedure of the various methods herein described. Since neither the dc field nor the permanent magnet are perfectly homogeneous, the best that can be accomplished will result in portions of the magnet being slightly magnetized in one direction and other portions in the opposite direction. DC fields are also applicable to calibration, although with less convenience than in the case of ac fields.

Thermal Method. If the temperature of a magnetized magnet is raised above the Curie temperature of the material and subsequently cooled to room temperature, it will be found that demagnetization has been achieved. Unfortunately, this is not practical for any of the metallic materials because heating these above the Curie point causes metallurgical changes resulting in unusably low magnetic properties thereafter. On the other hand, this is quite practical for any of the ceramic grades. The Curie temperature for these is approximately 460 °C, and a slight safety factor of 40 to 50 °C should be used to assure that all parts of the magnets will reach at least 460 °C. To avoid cracking because of heat shock, the ceramics should be heated and cooled relatively slowly, depending upon the size and shape of the part. Thermal demagnetization is not applicable to calibration.

5.5 Stability and Stabilization

Permanent magnets do not "run down." In this respect they differ fundamentally from batteries, radioactive materials, or the like. The magnetic field surrounding a magnet does not require energy to maintain it, therefore, there is no theoretical reason for a permanent magnet to continually lose strength. In practice, however, flux changes may occur as a result of several factors. Proper stabilization will eliminate or reduce these.

5.5.1 Metallurgical Changes

In the older permanent magnet materials, such as cobalt-steel, some metallurgical changes take place as a function of time. If such a magnet is magnetized before these changes have stabilized, flux changes superimposed on those to be described in the next section will occur. (This effect can be reduced to a negligible factor by artificial aging.) In the newer materials, such as Alnico or ceramic, metallurgical changes do not take place to any measurable degree at room temperature.

5.5.2 Time

A freshly magnetized permanent magnet will lose a minor percentage of its flux, as a function of time. It has been shown that, if one plots flux loss linearly against time logarithmically, an essentially straight line results. Laboratory measurements on some materials are shown in Table 5-2. All losses are based on measurements made starting at 0.1 after magnetizing.

TABLE 5-2 Laboratory measurements on selected magnet materials (percent).

Material	*Loss Per Log Cycle*	*Loss at 100,000 h (11.4 yrs)*
Ceramic	Essentially zero	Essentially zero
Alnico 3 (near max. energy)	0.4	2.4
Alnico 3 (near coercive)	0.6	3.6
Alnico 5 (near residual)	0.01	0.06
Alnico 5 (near max. energy)	0.15	0.9
Alnico 5 (near coercive)	0.4	2.4
Alnico 8—no data (expected to be less than Alnico 5)		

This loss in flux can be essentially eliminated by a partial demagnetization of the freshly charged magnet, in an amount of 7 to 15%. This is most conveniently accomplished by means of an ac coil. The ac field should be in the same direction as was the magnetizing field. It should be reduced to zero gradually, either by withdrawing the magnet with power applied or by reducing the ac voltage to zero while using a variable auto transformer.

5.5.3 Temperature

Temperature effects fall into three categories:

1. *Metallurgical changes* may be caused by exposure to too high a temperature. Such flux changes are not recoverable by remagnetization. The approximate maximum temperatures that can be used without experiencing metallurgical changes range from 550 °C for Alnico 5 to 1080 °C for the ceramics. The effect of metallurgical changes, if present, can be

avoided only by long-time exposure of the magnet to the temperature involved, prior to magnetizing.

2. *Irreversible losses* are defined as a partial demagnetization of the magnet, caused by exposure to high or low temperatures. Such losses are recoverable by remagnetization. Merely as examples, Table 5-3 shows values measured on laboratory specimens, with percent flux losses measured at room temperature after exposure to the indicated temperatures. Percentages shown in Table 5-3 are not additive for consecutive cycles above and below room temperature.

TABLE 5-3 Temperature effects on selected magnet materials (percent).

Material	*350°C*	*200°C*	*−20°C*	*−60°C*
Ceramic 5 $B_d/H_d > 2$ (above maximum energy)	0	0	0	0
Ceramic 6 $B_d/H_d > 1.1$ (near maximum energy)	0	0	0	0
Alnico 5 (near maximum energy)	1.3	0.8	1	2.5
Alnico 6 (near maximum energy)	0.6	0.4	0.5	1.3
Alnico 8 (near maximum energy)	0.3	0.2	0.1	0.1
Alnico 8 (near coercive)	3.5	2.0	0.5	0.8

The ideal method for stabilizing magnets against temperature-induced irreversible losses is installing them in the magnetic circuit for which they are intended, magnetize, then subject the assemblies to several temperature cycles which they are expected to experience in service. However, this is a time-consuming procedure that is normally impractical in production. Alternatively, the magnetized assembly may be partially demagnetized by means of an ac field, following the procedure described in the last paragraph of Sec. 5.5.2. A rule of thumb to follow is: determining by experiment that temperature cycling will cause *X%* flux loss, the ac field should be such as to cause a *2X%* flux loss, to properly stabilize against temperature.

3. *Reversible losses* are changes in flux that are reversible with temperature. For example, if any of the ceramic grades are heated 1 °C above room temperature, they will lose 0.19% of room temperature flux. However, this will be spontaneously regained upon the magnet cooling back at room temperature. The Alnico and E.S.D. materials have reversible variations on the order of 1/10 as great as the ceramics, depending upon the material and the operating point on the demagnetization curve. One cannot eliminate these reversible variations by stabilization treatments. However, use of proper temperature-compensation material in parallel with the magnet will reduce the effect to a negligible factor. Among others,

household watt-hour meter magnets and speedometer magnets are temperature-compensated in this manner.

5.5.4 Reluctance Changes

If a magnet is magnetized in a magnetic circuit and subsequently subjected to permeance change (such as changes in air gap dimensions or open-circuiting of the magnet) it may be found that a partial demagnetization of the magnet has occurred. Whether or not such a loss is experienced depends upon material properties and upon the extent of the permeance change.

Stabilization against such change is accomplished either by several times subjecting the magnet to such reluctance changes after magnetizing, or by use of the previously described ac field.

In this section it should be mentioned that contacting the magnet with ferromagnetic material (screwdrivers, pliers, and the like), at points other than the poles can cause an appreciable drop in flux at the poles. It is difficult to stabilize against this type of abuse. The remedy is to avoid such practices.

5.5.5 Adverse Fields

If a magnet or magnet assembly is subjected to an adverse magnetic field, a partial demagnetization may result, depending upon material properties and upon the intensity and direction of the adverse field. Proper stabilization consists of subjecting the magnet or assembly to a dc or ac demagnetizing field of the same magnitude as it is expected to encounter in service. The direction should be the same as that of the anticipated demagnetizing field.

5.5.6 Shock, Stress, and Vibration

The effects of shock, stress, and vibration below destructive limits on most permanent magnet materials are so minor (a few tenths of a percent) that little consideration need be given to them. Proper stabilization, as described in any of the preceding sections, will also stabilize against shock and vibration.

5.6 Measurements and Calibration

The selection of an appropriate test method for a permanent magnet should be based on full knowledge of its application. Information should establish the adverse conditions to which the magnet will be subjected in both the normal functioning and anticipated abuse of the product. Such information is frequently based on tests made on the product itself and should determine the flux density in the magnet, demagnetizing influences from counter magnetomotive force or variable permeance conditions, temperature limits, shock and vibration requirements, and other factors.

Properly correlated magnet tests require that the magnet be magnetized to saturation, in the proper direction, before proceeding with tests. It is imperative that the magnet be properly saturated, since improper magnetization frequently causes difficulty in obtaining correlation between magnetic tests and product performance (see Sec. 5.5.4).

Permanent-magnet acceptance tests as agreed upon by customer and supplier are generally based on tests and information obtained on the final product or inherent magnetic properties of the magnet.

Magnetic tests on permanent magnets can be either quantitative or empirical. Quantitative tests include measuring of absolute values of flux, flux density, force, and so on, in the end-use product. Empirical tests include arbitrary tests with magnetimeters, fluxmeters, gaussmeters such as Hall-effect devices, and other comparison devices. In most cases, quantitative test data are essential in setting up magnetic test specifications. Once the minimum standard and the proper test method have been established, the choice of test will depend largely upon the available equipment.

Although vital to the proper performance of the finished product and to the economic considerations involved, the design of the permanent magnet and its associated magnetic circuit is a problem that is fully discussed in available texts and design manuals.

It is assumed, therefore, that the design is adequate as applied to the final product and that the proper grade of permanent-magnet material has been selected and all dimensional factors have been optimized.

5.7 Handling of Permanent Magnets

The full utilization of permanent magnets is conditioned by proper handling prior to and during final usage. The natural phenomena of permanent magnets is such that, in many instances, the permanent magnet remembers what was done to it or what environment it was exposed to. Adverse factors affecting permanent magnets must be acknowledged and techniques applied to minimize or eliminate these deleterious conditions. The design, manufacture, handling, and processing of a permanent magnet is based on adequate control of this sequence to ensure optimum performance in the final application. It is no understatement that in a number of cases improper handling has resulted in poor, or completely substandard, performance.

A manufacturer who purchases and uses permanent magnets in a product has a very simple choice in the procurement of this essential material. The permanent magnet is purchased magnetized or unmagnetized. The initial choice may be simple, but either of the two conditions may result in quite distinct problems of processing the final device.

5.7.1 Unmagnetized Magnets

From the standpoint of ease in shipping, handling, and storage, the unmagnetized magnet is preferred. This condition alleviates special problems of knockdown, iron-chip pickup, and special instructions to manufacturing personnel.

To ensure magnetic quality, the permanent-magnet supplier has to test the magnet in a fully saturated magnetized condition. After this inspection the magnet is demagnetized prior to shipment. This implies that although the permanent-magnet material is purchased unmagnetized, it has been subjected to a complete magnetic cycle from an initial unmagnetized condition, through complete saturation, and back to an unmagnetized state. It is quite important to recognize that this sequence of events is a normal part of the inspection process for the permanent-magnet supplier.

The accrued benefits of purchasing, shipping, and processing of unmagnetized magnets are many. A few of the problem areas that can be eliminated or minimized for a permanent magnet in this unmagnetized state are:

Keepers are not required.

Proximity effects of other permanent magnets may be neglected.

Proximity of strong ac or dc fields pose no problem.

Physical shock or vibration for critical applications may be ignored.

Shape problems of self-demagnetization are not applicable.

Physical handling problems are considerably alleviated.

Storage and shipment problems are minimized.

A final step must be performed on this unmagnetized condition of the permanent magnet—magnetizing and inspecting the functional magnetic field of the permanent magnet after incorporating it in its intended device. This implies that the manufacturer of the device has adequate means for saturation and control of the magnetizing process for the permanent magnet. In some cases stabilization must be included for temperature, ac or dc electric fields, and/or other effects.

5.8 Glossary of Terms

Although there may be slight redundancy between this glossary of terms and the one in Appendix B, it was felt that for convenience it would be proper to include this section. Many of the terms that apply to magnetized core materials are not defined elsewhere in this volume—hence this important inclusion.

A_g, **Area of the Air Gap:** The cross-sectional area of the air gap perpendicular to the flux path, is the average cross-sectional area of that portion of the air gap within which the application interaction occurs. Area is measured in cm^2 in a plane normal to the central flux line of the air gap.

A_m, **Area of the Magnet:** The cross-sectional area of the magnet perpendicular to the central flux line, measured in cm^2 at any point along its length. In design, A_m is usually considered the area at the neutral section of the magnet.

B, **Magnetic Induction:** The magnetic field induced by a field strength, *H*, at a given point. It is the vector sum, at each point within the substance, of the magnetic field strength and resultant intrinsic induction. Magnetic induction is the flux per unit area normal to the direction of the magnetic path.

B_d, **Remanent Induction:** Any magnetic induction that remains in a magnetic material after removal of an applied saturating magnetic field, H_s. (B_d is the magnetic induction at any point on the demagnetization curve; measured in gauss.)

B_d/H_d, **Slope of the Operating Line:** The ratio of the magnetic induction, B_d, to its self-demagnetizing force, H_d. It is also referred to as the permeance coefficient, shear line, load line, and unit permeance.

B_dH_d, **Energy Product:** The energy that a magnet material can supply to an external magnetic circuit when operating at a point on its demagnetization curve; measured in megagauss-oersteds.

$(B_dH_d)_m$, **Maximum Energy Product:** The maximum energy a material can supply to an external magnetic circuit.

B_{is}, **Saturation Intrinsic Induction:** The maximum intrinsic induction possible in a material.

B_g, **Magnetic Induction in the Air Gap:** The average value of magnetic induction over the area of the air gap, A_{gi}; or it is the magnetic induction measured at a specific point within the air gap; measured in gauss.

B_1, **Intrinsic Induction:** The contribution of the magnetic material to the total magnetic induction, *B*. It is the vector difference between the magnetic induction in the material and the magnetic induction that would exist in a vacuum under the same field strength, *H*. This relation is expressed by the equation.

$$\mathbf{B}_1 = \mathbf{B} - \mathbf{H}$$

where $\mathbf{B}_1$ is the intrinsic induction in gauss, $\mathbf{B}$ the magnetic induction in gauss, and $\mathbf{H}$ the field strength in oersteds.

B_m, **Remanent Induction:** The magnetic induction that remains in a magnetic material after magnetizing and conditioning for final use; measured in gauss.

$B_m H_m$, **Energy Product:** The energy that a magnet material can supply to an external magnetic circuit when operating at the point $B_m H_m$; measured in megagauss-oersteds.

B_0, **Magnetic Induction:** The point of the maximum energy product $(B_d H_d)_m$; measured in gauss.

B_r, **Residual Induction (or Flux Density):** The magnetic induction corresponding to zero magnetizing force in a magnetic material after saturation in a closed circuit; measured in gauss.

B_s, **Saturation Induction:** The induction at saturation; measured in gauss.

f, **Reluctance Factor:** Accounts for the apparent magnetic circuit reluctance. This factor is required due to the treatment of H_m and H_g as constants.

F, **Leakage Factor:** Accounts for flux leakage from the magnetic curcuit. It is the ratio between the magnetic flux at the magnet neutral section and the average flux present in the air gap. $F = B_m A_m / B_g A_g$.

F, Magnetomotive Force (Magnetic Potential Difference): The line integral of the field strength, *H*, between any two points, p_1 and p_2.

$$F = \int_{P_1}^{P_2} H dl$$

where F is the magnetomotive force in gilberts, *H* the field strength in oersteds, and *dl* the element of length between the two points in centimeters.

H, **Magnetic Field Strength (Magnetizing or Demagnetizing Force):** The measure of the vector magnetic quantity that determines the ability of an electric current, or a magnetic body, to induce a magnetic field at a given point; measured in oersteds.

H_c, **Coercive Force, of a Material:** The demagnetizing force corresponding to zero magnetic induction, *B*, in a magnetic material after saturation; measured in oersteds.

H_{ci}, **Intrinsic Coercive Force, of a Material:** Its resistance to demagnetization. It is the demagnetizing force corresponding to zero intrinsic induction in a magnetic material after saturation; measured in oersteds.

H_d: That value of *H* corresponding to the remanent induction, B_d; measured in oersteds.

H_m: That value of *H* corresponding to the remanent induction, B_m; measured in oersteds.

H_0: The magnetic field strength at the point of the maximum energy product $(B_d H_d)_m$; measured in oersteds.

H_s, **Net Effective Magnetizing Force:** The magnetizing force required in the material, for saturation; measured in oersteds.

l_g, **Length of the Air Gap:** The length of the path of the central flux line of the air gap; measured in centimeters.

l_m, **Length of the Magnet:** The total length of magnet material traversed in one complete revolution of the center line of the magnetic circuit; measured in centimeters.

l_m/D, **Dimension Ratio:** The ratio of the length of a magnet to the diameter of a circle of equivalent cross-sectional area. For simple geometries, such as bars and rods, the dimension ratio is related to the slope of the operating line of the magnet, B_d/H_d.

P, **Permeance:** The reciprocal of the reluctance, p; measured in maxwells per gilbert.

R, **Reluctance:** Somewhat analogous to electrical resistance. It is the quantity that determines the magnetic flux, ϕ resulting from a given magnetomotive force, F.

$$R = \frac{F}{\phi}$$

where R is the reluctance in gilberts per maxwell, F the magnetomotive force in gilberts, and ϕ the flux in maxwells.

T_c, **Curie Temperature:** The transition temperature above which a material loses its permanent-magnet properties.

V_g, **Air-gap Volume:** The useful volume of air or nonmagnetic material between magnetic poles; measured in cubic centimeters.

μ, **Permeability:** A general term used to express various relationships between magnetic induction, B, and the field strength, H.

μ_{re}, **Recoil Permeability:** The average slope of the recoil hysteresis loop. (Also known as a minor loop.)

ϕ, **Magnetic Flux:** A contrived but measurable concept that has evolved in an attempt to describe the "flow" of a magnetic field. Mathematically, it is the surface integral of the normal component of the magnetic induction, B, over an area, A.

$$\phi = \int\int B\, dA$$

where ϕ is the magnetic flux in maxwells, B the magnetic induction in gauss, and dA the element of area in cm^2. When the magnetic induction,

B, is uniformly distributed and is normal to the area, *A*, the flux, ϕ, = *BA*.

A **Closed-Circuit Condition:** Exists when the external flux path of a permanent magnet is confined within high-permeability material.

A **Demagnetization Curve:** The second (or fourth) quadrant of a major hysteresis loop. Points on this curve are designated by the coordinates B_d and H_d.

A **Fluxmeter:** A galvanometer that measures the change of flux linkage with a search coil.

The **Gauss:** The unit of magnetic induction, *B*, in the cgs electromagnetic system. One gauss is equal to 1 maxwell per square centimeter.

A **Gaussmeter:** An instrument that measures the instantaneous value of magnetic induction, *B*. Its principle of operation is usually based on one of the following: the Hall effect, nuclear magnetic resonance (NMR), or the rotating coil principle.

The **Gilbert:** The unit of magnetomotive force, F, in the cgs electromagnetic system.

A **Hysteresis Loop:** A closed curve obtained for a material by plotting (usually to rectangular coordinates) corresponding values of magnetic induction, *B*, for ordinates and magnetizing force, *H*, for abscissa when the material is passing through a complete cycle between definite limits of either magnetizing force, *H*, or magnetic induction, *B*.

A **Keeper:** A piece (or pieces) of soft iron that is placed on or between the pole faces of a permanent magnet to decrease the reluctance of the air gap and thereby reduce the flux leakage from the magnet. It also makes the magnet less susceptible to demagnetizing influences.

Leakage Flux: Flux, ϕ, whose path is outside the useful or intended magnetic circuit; measured in maxwells.

The **Major Hysteresis Loop of a Material:** The closed loop obtained when the material is cycled between positive and negative saturation.

The **North Pole of a Magnet:** The pole that is attracted by the geographical North Pole. The north pole of a magnet repels the north-seeking pole of a compass.

The **Maxwell:** The unit of magnetic flux in the cgs electromagnetic system. One maxwell is one line of magnetic flux.

The **Neutral Section of a Permanent Magnet:** Defined by a plane passing through the magnet perpendicular to its central flux line at the point of maximum flux.

The **Oersted:** The unit of magnetic field strength, *H*, in the cgs electromagnetic system. One oersted equals a magnetomotive force of 1 gilbert per centimeter of flux path.

An **Open-Circuit Condition:** Exists when a magnetized magnet is by itself with no external flux path of high-permeability material.

The **Operating Line for a Given Permanent Magnet Circuit:** A straight line passing through the origin of the demagnetization curve with a slope of negative B_d/H_d. (Also known as permeance coefficient line.)

The **Operating Point of a Permanent Magnet:** That point on a demagnetization curve defined by the coordinates B_dH_d or that point within the demagnetization curve defined by the coordinates B_mH_m.

An **Oriented (Anisotropic) Material:** One that has better magnetic properties in a given direction.

A **Permanent Magnet:** A body that is capable of maintaining a magnetic field at other than cryogenic temperatures with no expenditure of power.

A **Permeameter:** A complex piece of equipment that can measure, and often record, the complete magnetic characteristics of a specimen of a magnetic material.

Magnetic Saturation of a Material: Exists when an increase in magnetizing force, H, does not cause an increase in the intrinsic magnetic induction, B, of the material.

A **Search Coil:** A coiled conductor, usually of known area and number of turns, that is used with a fluxmeter to measure the change of flux linkage with the coil.

The **Temperature Coefficient:** A number that describes the change in a magnetic property with a change in temperature. It usually is expressed as the percentage change per unit of temperature.

An **Unoriented (Isotropic) Material:** Equal magnetic properties in all directions.

BIBLIOGRAPHY

BARTA, G.T., "Permanent Magnet Processing," *Electro-Technology,* May 1966.

CLEGG, A.G., AND M. MCCRAIG, "High-Temperature Stability of Permanent Magnets of the Iron–Nickel–Aluminum System," *British J. Appl. Phys.,* Vol. 9, No. 144, 1958.

DIETRICH, H., "Time and Temperature Dependence of the Structural Changes in Permanent Magnet Materials," *Cobalt,* No. 30, March 1966, pp. 3–18.

Gordon, D.I., "Environmental Evaluation of Magnetic Materials," *Electro-Technology,* Vol. 67, No. 1, January 1961, pp. 118–125.

Gordon, D.I., "Irradiating Magnetic Materials," *Electro-Technology,* June 1965, pp. 42–45.

Gordon, D.I., and R.S. Sery, *IEEE Trans. Commun. Electron.,* Vol. CE-83, No. 73, July 1964, pp. 357–361.

Gordon, D.I., R.S. Sery, and R.H. Lundsten, "Nuclear Radiation Effects in Magnetic Core Materials and Permanent Magnets," ONR-5, Materials Research in the Navy, Symposium, Philadelphia, 1959 (Office of Naval Research, Washington, D.C., 1959), pp. 253–292.

Hadfield, D., *Permanent Magnets and Magnetism.* John Wiley & Sons, Inc., New York, 1962.

Kronenberg, K.J., and M.A. Bohlmann, "Long Term Magnetic Stability of Alnico V and Other Permanent Magnet Materials," *WADC Rept. 58-535.* "Long Term Magnetic Stability of Alnico and Barium Ferrite Magnets," *J. Appl. Phys.,* Vol. 31, May 1960, pp. 825–845.

Parker, R.J. and R.J. Studders, *Permanent Magnets and Their Application.* John Wiley & Sons, Inc., New York, 1962.

Roberts, W.H., "Performance of Permanent Magnets at Elevated Temperatures," *J. Appl. Phys.,* Vol. 29, No. 405, 1958.

Schindler, A.I., and E.I. Salfovitz, "Effect of Applying a Magnetic Field during Neutron Irradiation on the Magnetic Properties of Fe–Ni Alloys," *J. Appl. Phys.,* Vol. 31 supplement to No. 5, 1960, p. 245–S.

Sery, R.S., and D.I. Gordon, "Irradiation of Magnetic Materials with 1.5- and 4-meV Protons," *J. Appl. Phys.,* Vol. 67, No. 4 (Pt. 2), April 1963, pp. 1311–1312.

Sery, R.S., D.I. Gordon, and R.H. Lundsten, NAVORD Report 6276 (U.S. Naval Ordnance Laboratory, White Oak, Md.), 1959.

Sery, R.S., R.H. Lundsten, and D.I. Gordon, Naval Ordnance Laboratory TR-61-45, Silver Spring, Md., May 18, 1961.

Spreadbury, F.G., *Permanent Magnets.* Pitman and Sons Ltd., London, 1949.

Tenzer, R.K., "Influence of Various Heat Exposures of Alnico V Magnets," *J. Appl. Phys.,* Vol. 30, supplement to No. 4, 1959, p. 115–S.

Tenzer, R.K., "Temperature Effects of the Remanence of Permanent Magnets," Technical Documentary Report ASD-TDR-63-500, Office of Technical Services AD-420235, U.S. Department of Commerce.

APPENDIX A

REFERENCES

No book can be considered complete without the inclusion of a comprehensive list of selected papers, manufacturers' application notes, and books. Although there are a number of references that relate to magnetic-core materials and their applications, only those which the author feels are absolutely pertinent to the subject are listed here. This bibliography provides an exceptionally concise cross section of the available material which can be used to broaden the reader's knowledge of magnetic materials and how they can be applied in practical experience.

The listing that follows immediately is provided by the IEC. It will be of particular interest to those who plan to, or are presently engaged in commercial design and manufacture of magnetic-core components. These documents can be obtained from the American National Standards Institute, Inc., 1430 Broadway, New York, NY 10018.

The following compilation of selected papers is offered as reference material on a wide spectrum of subjects that deal with magnetic cores and their applications. The material is useful to the engineer, technician, or student as supplemental theory to that provided in this book.

SELECTED TECHNICAL PAPERS

[1] **B. Astle**, "Optimum Shapes for Inductors," *IEEE Trans. PM & P,* March 1969.

[2] **H. Blinchikoff**, "Toroidal Inductor Design," *Electro-Technology,* Nov. 1964.

[3] **Burnell & Co. Engineering Staff**, "The Application of Iron Powder Cores in Electric Wave Filters," *Progr. Powder Metall.,* 1962.

TABLE A-1 IEC publications on linear ferrites.

125 (1961)	General classification of ferromagnetic oxide materials and definitions of terms. Amendment No. 1 (1965) Amendment No. 2 (1968)
133 (1967)	Dimensions for pot cores made of ferromagnetic oxides and associated parts (2nd edition).
133A (1970)	First supplement to Publication 133.
133B (1971)	Second supplement to Publication 133. Amendment No. 1 (1975)
205 (1966)	Calculation of the effective parameters of magnetic piece parts.
205A (1968)	Supplement to Publication 205.
205B (1974)	Supplement to Publication 205.
220 (1966)	Dimensions of tubes, pins, and rods of ferromagnetic oxides.
221 (1966)	Dimensions of screw cores made of ferromagnetic oxides. Amendment No. 1 (1968) Amendment No. 2 (1976)
221A (1972)	First supplement to Publication 221.
223 (1966)	Dimensions of aerial rods and slabs of ferromagnetic oxides.
223A (1972)	First supplement to Publication 223.
226 (1967)	Dimensions of cross cores (X-cores) made of ferromagnetic oxides and associated parts.
226A (1970)	First supplement to Publication 226.
367-1 (1971)	Cores for inductors and transformers for telecommunication. Part 1—Measuring methods. Amendment No. 1 (1976)
367-1A (1973)	First supplement.
367-1B (1973)	Second supplement to Publication 367-1.
367-1C (1974)	Third supplement.
367-2 (1974)	Part 2—Guides for the drafting of performance specifications.
367-2A (1976)	First supplement to Publication 367-2.
401 (1972)	Information on ferrite materials appearing in manufacturers' catalogues of transformer and inductor cores.
424 (1973)	Guide to the specification of limits for physical imperfections of parts made from magnetic oxides.
431 (1973)	Dimensions of square cores (RM cores) made of magnetic oxides and associated parts.
525 (1976)	Dimensions of toroids made from magnetic oxides or iron powders.

[4] **E.H. Chant, Jr.**, "Moly Permalloy Cores," *MPIF Sem.,* June 1969.

[5] **R.A. Chewidden**, "A Review of Magnetic Materials—Especially for Communications Systems," *Metal Progr.,* 1948.

[6] **E. G. Cristal**, "Tables of Maximally Flat Impedance-Transforming Networks of Low-Pass-Filter Form," *IEEE Trans. Microwave Theory Techniques,* Vol. MTT 13, No. 5, Sept. 1965, Corres.

[7] **M.F. "Doug" DeMaw**, "The Practical Side of Toroids," *QST Magazine* (ARRL, Inc.), June 1979.

[8] **R.M. Fano**, "Theoretical Limitations on the Broadband Matching of Arbitrary Impedances," *J. Franklin Inst.,* January–February 1950.

[9] **G.B. Finke**, "Moly Permalloy Powder Cores—Their Characteristics and Applications," *MPIF Sem.,* December 1969.

[10] **E.A. Gaugler**, **"Soft Magnetic Materials,"** *Product Eng.,* July 1949.

[11] **A.J. Harendza-Harinxma**, "Recent Developments in the Manufacture of Permalloy Powder Cores," *Western Electric Eng.,* 1964.

[12] **J.H. Horowitz**, "Design Wideband UHF Power Amplifiers," *Electron. Design,* May 24, 1969.

[13] **G.A. Kelsall**, "Permeameter for Alternating Current Measurements at Small Magnetizing Forces," *Opt. Soc. Amer. J.,* Vol. 8, Feb. 1924.

[14] **Bohdan Kostyshyn and Peter H. Haas**, "Discussion of Current-Sheet Approximations in Reference to High-Frequency Magnetic Measurements," *J. Res. NBS,* Vol. 52, No. 6, June 1954.

[15] **Krauss-Allen**, "Designing Toroidal Transformers to Optimize Wideband Performance," *Electronics,* Aug. 1973.

[16] **Lefferson**, "Twisted-Wire Transmission Line," *IEEE Trans. Parts, Hybrids, Packaging.* Vol. PHP-7, No. 4, Dec. 1971.

[17] **V.E. Legg**, "Magnetic Measurements at Low Flux Densities Using the A.C. Bridge," *Bell Systems Tech. J.,* Vol. 15, Jan. 1936.

[18] **V.E. Legg**, "Analysis of Quality Factor of Annular Core Inductors," *Bell System Tech. J.,* Vol. 39, No. 1, Jan. 1960.

[19] **V.E. Legg and F.J. Given**, "Compressed Powdered Molybdenum Permalloy for High Quality Inductance Coils," *Bell System Tech. J.,* Vol. 19, July 1940.

[20] **G.L. Matthaei**, "Tables of Chebyshev Impedance-Transforming Networks of Low-Pass Filter Form," *Proc. IEEE,* Aug. 1964.

[21] **E.J. Oelberman**, "Moisture Aging of Powder Core Toroids," *Electronics,* May 1953 (with R.E. Skipper and W.J. Leiss).

[22] **C.D. Owens**, "Stability Characteristics of Molybdenum Permalloy Powder Cores," *Elec. Eng.,* March 1956.

[23] **O. Pitzalis and R.A. Gilson**, "Tables of Impedance Matching Networks Which Approximate Prescribed Attenuation versus Frequency Slopes," *IEEE Trans. Microwave Theory Techniques,* Vol. MTT-19, No. 4, April 1971.

[24] **O. Pitzalis and Couse**, "Broadband Transformer Design for RF Transistor Amplifiers," *ECOM-2989,* U.S. Army Electronics Command, Ft. Monmouth, N.J., July 1968.

[25] **Stanley Pro**, "Toroid Design Analysis," *Electro-Technology,* Aug. 1966.

[26] **W. Quefurth**, "Turns Computation for Small Core Inductances," *Elec. Design News,* Sept. 1958.

[27] **C.E. Richards, P.R. Bardell, S.E. Buckley, and A.C. Lynch**, "Some Properties and Tests of Magnetic Powders and Powder Cores," *Elec. Commun.,* Vol. 28, March 1951.

[28] **C.L. Ruthroff**, "Some Broadband Transformers," *Proc. IRE,* Vol. 47, August 1959.

[29] **C.F. Salt, W.T. Sackett, Jr., and R.C. McMaster**, "Research and Development of Various Configurations of Core Materials for Optimum Transformer Design," Battelle Memorial Institute, Columbus, Ohio, 1952.

[30] **J. Sevick**, "Simple Broadband Matching Networks," *QST Magazine* (ARRL, Inc.), January 1976.

[31] **M.D. Sutton**, "Application of Iron Powder Cores in electromagnetic Delay Lines," *Progr. Powder Metall.,* 1962.

[32] **R. Turrin**, "Application of Broadband Balun Transformers," *QST Magazine* (ARRL, Inc.), April 1969.

[33] **J.E. Wolf and B. Kramer**, "A Production Technique for the Determination of Inductance for Toroidal Powdered Iron Cores," *Prog. Powder Metall.,* 1964.

[34] **T.G. Wilson, S.Y. Feng, and W.A. Sander**, "Optimum Toroidal Inductor Design Analysis," *Proc. 1970 Electron. Components Conf.*

[35] Symposium of Papers on Ferromagnetic Materials, IEE, Vol. 97, Part II, No. 56, April 1950.

APPLICATION NOTES

Manufacturers' application notes are generally available at no cost to those who request them by mail. Complete lists of each manufacturer's technical bulletins and application notes are also available upon request. The names and addresses of various manufacturers of magnetic-core materials are given following the list of notes.

[1] Broadband Linear Power Amplifiers Using Push-Pull Transistors," by H. Granberg, Motorola Semiconductor Products, Inc., *AN–593.*

[2] Design of H.F. Wideband Power Transformers," by Hilbers, *Phillips Application Information No. 530.*

[3] "Get 300 Watts PEP Linear Across 2 to 30 MHz from this Push-Pull Amplifier," by H. Granberg, Motorola Semiconductor Products, Inc., *EB–27.*

[4] "Matching Network Designs with Computer Solutions," Motorola Semiconductor Products, Inc., *AN-267.*

[5] "Notes on Low Impedance H.F. Broadband Transformer Techniques," by Lewis, Collins Radio Co., November 1964.

[6] "Precision Toroidal Inductors," *Sprague Tech. Bull. No. 41000.*

[7] "Systemizing RF Power Amplifier Design," Motorola Semiconductor Products, Inc., *AN-282.*

MANUFACTURERS

Amidon Associates, 12033 Otsego St., North Hollywood, CA 91607.

Arnold Engineering Co., Marengo, IL 60152.

Fair-Rite Products Corp., Walkill, NY 12589.

Ferroxcube Corp., 5083 Kings Highway, Saugerties, NY 12477.

Indiana General, Crows Mill Road, Keasbey, NJ 08832.

Magnetics (Division of Spang Industries), P.O. Box 391, Butler, PA 16001.

Micrometals, Inc., 228 North Sunset, City of Industry, CA 91744.

Motorola Semiconductor Products, Inc., P.O. Box 20912, Phoenix, AZ 85036.

PERTINENT BOOKS

The subjects of magnetic core materials and related network design are treated in the following books. The engineer or student may find these references of value as supplemental material to the topics covered in this volume.

[1] **P.R. Bardell**, *Magnetic Materials in the Electrical Industry.* Philosophical Library, Inc., New York, 1955.

[2] **H.W. Bode**, *Network Analysis and Feedback Amplifier Design.* D. Van Nostrand Co., New York.

[3] **R.M. Bozorth**, *Ferromagnetism.* D. Van Nostrand Co., New York, 1951.

[4] **F.C. Connelly**, *Transformers.* Pitman and Sons Ltd., London, 1950.

[5] **M.F. "Doug" DeMaw**, *Practical RF Communications Data for Engineers and Technicians* (No. 21557). Howard W. Sams & Co., Indianapolis, Ind., 1978.

(6) **M.F. "Doug" DeMaw**, *ARRL Electronics Data Book.* American Radio Relay League, Inc., Newington, Conn., 1976.

[7] **P.R. Geffe**, *Simplified Modern Filter Design.* Hayden Book Co., Inc., New York.

[8] **R. Lee**, *Electronic Transformers and Circuits.* John Wiley & Sons, Inc., New York, 1955.

[9] DEPARTMENT OF ELECTRICAL ENGINEERING, M.I.T., *Magnetic Circuits and Transformers.* John Wiley & Sons, Inc., New York, 1943.

[10] **W.J. POLYDOROFF**, *High-Frequency Magnetic Materials.* John Wiley & Sons, Inc., New York, 1960.

[11] H.F. STORM, *Magnetic Amplifiers.* John Wiley & Sons, Inc., New York, 1955.

APPENDIX **B**

MISCELLANEOUS DATA

This appendix contains an assortment of useful information for engineers and students who are involved with ferromagnetic components. The author has endeavored to include a wide assortment of data that would otherwise be found scattered throughout the countless technical literature which pertains to components that contain magnetic-core material. Credits are given wherever possible to those manufacturers and distributors of magnetic cores from which the information has been gleaned.

Table B-1 contains a comprehensive listing of the symbols ánd their definitions for magnetic-core materials. These symbols and terms are standard within the industry, with the notable exception of the expressions of permeability. Some manufacturers designate initial permeability as μ_0, whereas others assign the symbol μ_i. Similarly, some fabricators prefer μ_{av} to μ_e when symbolizing average or effective permeability.

SELECTED CONVERSION FACTORS

Table B-2 contains a group of important conversion factors which find frequent application among engineers and students who are concerned with the design and use of components that utilize magnetic cores. Metric/English conversion data are also included for those areas of engineering that encompass transformer and inductor design.

TABLE B-1 Magnetic-core symbology.

Symbol	*Units*	*Term*	*Description*
A_c	in^2 cm^2	Available winding area of core	Cross-sectional area (perpendicular to direction of wire), available for winding turns on a particular core.
A_{CB}	in^2 cm^2	Available winding area of bobbin	Cross-sectional area (perpendicular to direction of wire), available for winding turns on a particular bobbin.
A_e	cm^2	Effective area of core	The cross-sectional area that an equivalent gapless core (of uniform magnetic and geometric properties) would have.
A_g	cm^2	Effective gap area	The equivalent area through which the (assumed uniform) flux in a magnetic core gap passes. Corrects for fringing effects.
A_L	mH/1000 turns	Inductance index	Relates inductance to turns for a particular core and gap.
A_m	cm^2	Effective area of magnetic path	In a gapped structure, the equivalent cross-sectional area of the magnetic part of the path, assumed uniform.
A_p	in^2 cm^2	Available winding space for primary	That portion of the total available winding area allotted to the transformer primary winding.
A_x	in^2 cm^2	Cross-sectional area of wire	The cross-sectional area of the conductor part of the wire.
B	gauss	Magnetic flux density	The flux density (lines/cm^2) in a magnetic circuit, measured at a given point.
B_{max}	gauss	Maximum flux density	The value of flux density corresponding to the peak of the applied excitation.
B_o	gauss	Magnetic flux density	Maximum values of flux density for $\mu = \mu_e$ in transformer core.
B_t	gauss	Pulse-excited flux density	The value of flux density corresponding to the instant of termination of an applied rectangular pulse.
BW	hertz	Bandwidth	Frequency range over which response of a tuned transformer is considered to be uniform $\left(BW = 2\Delta f = \frac{f_0}{Q}\right)$
C_t	farads	Equivalent shunt capacitance	The equivalent shunt capacitance represented by the interwinding capacitance of a transformer.
E_m	volts, RMS	Magnetizing voltage	Effective value of alternating voltage applied across primary magnetizing inductance (see L_m).
E_p	volts, RMS	Transformer primary terminal voltage	Effective value of alternating voltage applied across primary terminals of a transformer.
E_{pmax}	volts, Peak	Peak primary	Peak value of E_p (for sine waves, $E_{pmax} = E_p\sqrt{2}$).
E_s	volts, RMS	Transformer secondary terminal voltage	Effective value of alternating voltage applied across secondary (output) terminals of a transformer.
f	hertz	Frequency	Alternations per second of voltage or current.
f_1	hertz	Low-frequency cut-off	Lowest frequency at which the sinusoidal amplitude vs. frequency characteristic of a transformer is within 3 dB of the midfrequency leve

TABLE B-1 (continued)

ymbol	*Units*	*Term*	*Description*
	hertz	High-frequency cut-off	Highest frequency at which the sinusoidal amplitude vs. frequency characteristic of a transformer is within 3 dB of the midfrequency level.
	oersteds	Magnetization	The magnetizing force that produces the magnetic flux in a transformer or inductor core. Ampere-turns per cm.
nax	oersteds	Peak magnetization	The peak value of the magnetizing force.
	oersteds	Dc magnetization	The dc magnetizing force applied to a core (unidirectional current flowing in its windings).
	amperes, RMS	Core loss current	Component of magnetizing current accounting for power lost in core material.
	amperes	Direct current	The magnitude of the unidirectional component of a current.
	amperes, RMS	Magnetizing current	Component of primary alternating current devoted to magnetization of transformer core.
	amperes, RMS	Transformer primary current	Total alternating current flowing in transformer primary winding. Vector sum of magnetizing (l_m) and load ($l_s \div n$) currents.
nap	amperes, Peak	Peak input current	The peak value of the transformer input current.
	amperes, RMS	Transformer secondary current	Total alternating current flowing in transformer secondary winding.
	numeric	Dielectric constant of a substance	The ratio of the capacitance of a capacitor using that substance as a dielectric to the capacitance of the same structure with a vacuum as the dielectric.
	(gauss)^2	μ vs. B factor	Factor relating μ and B for a given power inductor design requirement.
	volt-amperes per $H_z \times 2\pi$	Core-selection factor	Factor relating the volt-ampere rating of a core to the frequency.
m	henries	Self-inductance	Self-inductance of a transformer primary winding ($L_m = N_p^2 AL \times 10^{-9}$ henry)
0	henries	Air inductance	L for $\mu = 1$ (air). Thus, $L = \mu_0 L_0$.
	henries	Leakage inductance	Self-inductance that appears in series with the windings due to leakage flux that does not contribute to the transfer of energy in a transformer.
	in. cm.	Length of flux path	Effective length of a uniform flux path.
	in. cm.	Length of air gap	Effective length of a uniform air gap in a magnetic circuit.
n	in. cm.	Length of magnetic path	Effective length of a uniform magnetic path.
	numeric	Load/magnetizing current ratio	The ratio of the load component ($l_s \div n$) to the magnetizing component (l_m) in the transformer primary current (l_p).
p	numeric	Primary turns	Total number of complete turns in transformer primary winding.
s	numeric	Secondary turns	Total number of complete turns in transformer secondary winding.
	numeric	Turns ratio	Ratio of transformer primary/secondary turns. Equivalent to primary/secondary voltage ratio (neglecting resistance of windings)

TABLE B-1 (continued)

Symbol	*Units*	*Term*	*Description*
P_d	percent	Droop	Percent droop in the nominally flat portion of the waveform of a pulse delivered by a transformer, compared with the input pulse wavefo
P_{load}	watts	Load power	The power dissipated in the external load connected to the transformer secondary termi
P_0	watts	Power dissipation factor	The power dissipation level that will cause a temperature rise of 50°C above ambient in a given core.
P_t	watts	Total power dissipation	The total power dissipated in an inductor (winding loss + core loss). Neglects dielectric losses.
Q	numeric	Quality factor	Ratio of reactance to equivalent series resistance in an impedance.
R_C	ohms	Equivalent core loss resistance	The value of resistance across which the magneti excitation voltage would dissipate a power equ to the core loss of a transformer or inductor.
R_g	ohms	Generator resistance	Effective internal series resistance of the excitation source.
R_L	ohms	Load resistance	Equivalent resistance of external load connected between transformer secondary termin
R_p	ohms	Resistance of primary winding	Dc resistance of transformer primary.
R_s	ohms	Resistance of secondary winding	Dc resistance of transformer secondary.
T_{max}	°C	Maximum operating temperature	The sum of the maximum expected ambient temperature and the winding temperature rise in the core material.
T_{rise}	°C	Temperature rise	Rise in temperature of an inductor or transformer core due to internal power dissipation.
t_p	seconds	Pulse duration	Duration of the longest nominally flat portion of a pulse waveform.
t_r	seconds	Pulse rise time	Longest allowable time for a pulse to rise from 10% to 90% of its final amplitude.
V_e	in.^{3} cm^3	Effective core volume	The effective volume of a magnetic core ($V_e = l_e \times A_e$) defined as the volume of a uniform magnetic path l_e long, having a cross sectional area A_e.
X_t	ohms	Leakage reactance	The reactance of transformer leakage inductance $L_t (X_t = 2\pi f L_t)$.
α	numeric	Attenuation constant	Midfrequency reduction in signal amplitude from primary terminal voltage to voltage established across the load.
$\alpha\ t$	numeric	Total attenuation	Midfrequency reduction in signal amplitude from generator internal EMF to transformer output terminal voltage.
Δf	hertz	Half-bandwidth	Difference between center frequency and low frequency cutoff point (f_1) on resonance curve of tuned transformer.
μ	numeric	Permeability	Ratio of magnetic flux density to the excitation producing it ($\mu = B/H$).
av	numeric	Average permeability	The slope of the straight line between 0, 0, and $B = B_{max}$ for a given core material ($\mu\ \text{av} = \frac{B_{max}}{H_{max}}$)
$\mu\ 0$ $\mu\ i$	numeric	Initial permeability	The effective permeability of the core at low excitation (in the linear region) for a uniform magnetic path.

Courtesy of Ferroxcube Corp.

TABLE B-2 Selected conversion factors.

Multiply:	*By:*	*To Obtain:*
	Weight	
pounds	453.59	grams
pounds	0.45359	kilograms
grams	0.0022046	pounds
kilograms	2.2046	pounds
	Length	
feet	30.480	centimeters
inches	2.5400	centimeters
centimeters	0.032808	feet
centimeters	0.39370	inches
inches	0.025400	meters
meters	39.370	inches
	Area	
square feet	929.04	square centimeters
square inches	6.4516	square centimeters
square centimeters	1.0764×10^{-3}	square feet
square centimeters	0.15500	square inches
square inches	6.4516×10^{-4}	square meters
square meters	1.5500×10^{3}	square inches
square centimeters	10^{-4}	square meters
square meters	10^{4}	square centimeters
	Sinusoidal Waveform	
peak current or voltage	0.70711	rms current or voltage
peak current or voltage	0.63662	average current or voltage
rms current or voltage	1.4142	peak current or voltage
rms current or voltage	0.90032	average current or voltage
average current or voltage	1.5708	peak current or voltage
average current or voltage	1.1107	rms current or voltage
	Magnetic Induction, *B*	
gauss	6.4516	lines per square inch
gauss	6.4516×10^{-8}	webers per square inch
gauss	10^{-4}	webers per square meter (teslas)
lines per square inch	0.15500	gausses
lines per square inch	1.5500×10^{-5}	webers per square meter (teslas)
lines per square inch	10^{-8}	webers per square inch

TABLE B-2 (continued)

Multiply:	*By:*	*To Obtain:*
webers per square inch	1.5500×10^7	gausses
webers per square inch	10^8	lines per square inch
webers per square inch	1550	webers per square meter (teslas)
	Magnetizing Force, H	
oersteds	2.0213	ampere-turns per inch
oersteds	0.79577	ampere-turns per centimeter
oersteds	79.577	ampere-turns per meter
ampere-turns per centimeter	1.2566	oersteds
ampere-turns per centimeter	2.5400	ampere-turns per inch
ampere-turns per centimeter	100.00	ampere-turns per meter
ampere-turns per inch	0.49474	oersteds
ampere-turns per inch	0.39370	ampere-turns per centimeter
ampere-turns per inch	39.370	ampere-turns per meter
ampere-turns per meter	0.012566	oersteds
ampere-turns per meter	10^{-2}	ampere-turns per centimeter
ampere-turns per meter	0.025400	ampere-turns per inch
	Permeability	
gauss per oersted	3.1918	lines per ampere-turn inch
gauss per oersted	3.1918×10^{-8}	webers per ampere-turn inch
gauss per oersted	1.2566×10^{-6}	webers per ampere-turn meter
webers per ampere-turn meter	7.9577×10^5	gauss per oersted
webers per ampere-turn meter	2.5400×10^6	lines per ampere-turn inch
webers per ampere-turn meter	0.025400	webers per ampere-turn inch
webers per ampere-turn inch	3.1330×10^7	gauss per oersted
webers per ampere-turn inch	10^8	lines per ampere-turn inch
webers per ampere-turn inch	39.370	webers per ampere-turn meter
lines per ampere-turn inch	0.31330	gauss per oersted
lines per ampere-turn inch	39.370×10^{-8}	webers per ampere-turn meter
lines per ampere-turn inch	10^{-8}	webers per ampere-turn inch

CONVERSION INFORMATION FOR WIRE GAUGES 10 TO 44

Columns A and B of Table B-3 provide data on the common circular mils notation and the metric equivalent for each wire gauge. Column C is useful to the engineer or student, as it provides the equivalent resistance in microhms/centimeters ($\mu\Omega$/cm or 10^{-6} Ω/cm). Columns D through L contain important information concerning coated wires and the effect of the insulation on the size and number of turns, plus the total weight in g/cm.

The designer can determine the total resistance of a winding by multiplying the MLT (mean length/turn) of the winding in centimeters by the $\mu\Omega$/cm for the chosen wire gauge and the total number of turns. Hence,

$$R = (\text{MLT}) \times \frac{\mu\Omega}{\text{cm}} \times N$$

where N is the number of turns employed.

The weight of the copper in a specified winding can be found by multiplying the MLT by the g/cm (see column L) and by the total number of turns.

$$\text{wt} = (\text{MLT}) \times \frac{\text{g}}{\text{cm}} \times N$$

The turns per square inch and per cm² are based on a 60% wire-fill factor.

INDUCTANCE, CAPACITANCE, AND REACTANCE

Although a calculator can be used to learn the inductance, capacitance, or reactance of circuit elements, it is sometimes more convenient to obtain a quick approximation by means of a nomograph. Figure B-1 contains the necessary information for those who prefer the nomograph format. Inductance values from 0.01 μH to 1000 H are provided. Reactance in ohms is included from 0.01 Ω to 1 MΩ. Capacitance values from 0.1 μF to 10 μF are listed and frequencies between 10 Hz and 1 MHz are included.

RESISTANCE/TEMPERATURE CORRECTION FACTOR

In many design applications it is imperative to know the resistance of a transformer or inductor winding with respect to t_{av} (average temperature). Correction factors provide suitable resistance values versus temperature for

TABLE B-3 Circular mils notation and metric equivalents.

AWG Wire Size	Bare Area: $cm^2 \times 10^{-3}$ [a]	Bare Area: cir. mil [b]	Resistance: 10^{-6}/ / cm at 20°C	Heavy Synthetics Area: $cm^2 \times 10^{-3}$	Heavy Synthetics Area: cir mil [b]	Diameter: cm	Diameter: in. [b]	Turns per: cm	Turns per: in. [b]	Turns per: cm^2	Turns per: $in.^2$	Weight: g/cm
10	52.61	10,384	32.70	55.9	11,046	0.267	0.1051	3.87	9.5	10.73	69.20	0.468
11	41.68	8,226	41.37	44.5	8,798	0.238	0.0938	4.36	10.7	13.48	89.95	0.3750
12	33.08	6,529	52.09	35.64	7,022	0.213	0.0838	4.85	11.9	16.81	108.4	0.2977
13	26.26	5,184	65.64	28.36	5,610	0.190	0.0749	5.47	13.4	21.15	136.4	0.2367
14	20.82	4,109	82.80	22.95	4,556	0.171	0.0675	6.04	14.8	26.14	168.6	0.1879
15	16.51	3,260	104.3	18.37	3,624	0.153	0.0602	6.77	16.6	32.66	210.6	0.1492
16	13.07	2,581	131.8	14.73	2,905	0.137	0.0539	7.32	18.6	40.73	262.7	0.1184
17	10.39	2,052	165.8	11.68	2,323	0.122	0.0482	8.18	20.8	51.36	331.2	0.0943
18	8.228	1,624	209.5	9.326	1,857	0.109	0.0431	9.13	23.2	64.33	414.9	0.07472
19	6.531	1,289	263.9	7.539	1,490	0.0980	0.0386	10.19	25.9	79.85	515.0	0.05940
20	5.188	1,024	332.3	6.065	1,197	0.0879	0.0346	11.37	28.9	98.93	638.1	0.04726
21	4.116	812.3	418.9	4.837	954.8	0.0785	0.0309	12.75	32.4	124.0	799.8	0.03757
22	3.243	640.1	531.4	3.857	761.7	0.0701	0.0276	14.25	36.2	155.5	1,003	0.02965
23	2.588	510.8	666.0	3.135	620.0	0.0632	0.0249	15.82	40.2	191.3	1,234	0.02372
24	2.047	404.0	842.1	2.514	497.3	0.0566	0.0223	17.63	44.8	238.6	1,539	0.01884
25	1.623	320.4	1,062.0	2.002	396.0	0.0505	0.0199	19.80	50.3	299.7	1,933	0.01498
26	1.280	252.8	1,345.0	1.603	316.8	0.0452	0.0178	22.12	56.2	374.2	2,414	0.01185
27	1.021	201.6	1,687.6	1.313	259.2	0.0409	0.0161	24.44	62.1	456.9	2,947	0.00945
28	0.8046	158.8	2,142.7	1.0515	207.3	0.0366	0.0144	27.32	69.4	570.6	3,680	0.00747
29	0.6470	127.7	2,664.3	0.8548	169.0	0.0330	0.0130	30.27	76.9	701.9	4,527	0.00602
30	0.5067	100.0	3,402.2	0.6785	134.5	0.0294	0.0116	33.93	86.2	884.3	5,703	0.00472
31	0.4013	79.21	4,294.6	0.5596	110.2	0.0267	0.0105	37.48	95.2	1,072	6,914	0.00372
32	0.3242	64.00	5,314.9	0.4559	90.25	0.0241	0.0095	41.45	105.3	1,316	8,488	0.00305
33	0.2554	50.41	6,748.6	0.3662	72.25	0.0216	0.0085	46.33	117.7	1,638	10,565	0.00241
34	0.2011	39.69	8,572.8	0.2863	56.25	0.0191	0.0075	52.48	133.3	2,095	13,512	0.00189
35	0.1589	31.36	10,849	0.2268	44.89	0.0170	0.0067	58.77	149.3	2,645	17,060	0.00150
36	0.1266	25.00	13,608	0.1813	36.00	0.0152	0.0060	65.62	166.7	3,309	21,343	0.00119
37	0.1026	20.25	16,801	0.1538	30.25	0.0140	0.0055	71.57	181.8	3,901	25,161	0.000977
38	0.08107	16.00	21,266	0.1207	24.01	0.0124	0.0049	80.35	204.1	4,971	32,062	0.000773
39	0.06207	12.25	27,775	0.0932	18.49	0.0109	0.0043	91.57	232.6	6,437	41,518	0.000593
40	0.04869	9.61	35,400	0.0723	14.44	0.0096	0.0038	103.6	263.2	8,298	53,522	0.000464
41	0.03972	7.84	43,405	0.0584	11.56	0.00863	0.0034	115.7	294.1	10,273	66,260	0.000379
42	0.03166	6.25	54,429	0.04558	9.00	0.00762	0.0030	131.2	333.3	13,163	84,901	0.000299
43	0.02452	4.84	70,308	0.03683	7.29	0.00685	0.0027	145.8	370.4	16,291	105,076	0.000233
44	A 0.0202	B 4.00	C 85,072	D 0.03165	E 6.25	F 0.00635	G 0.0025	H 157.4	I 400.0	J 18,957	K 122,272	L 0.000195

[a] This notation means that the entry in the column must be multiplied by 10^{-3}.
[b] These data from *REA Magnetic Wire Dataloator.*
Courtesy of Magnetics, Division of Spang Industries.

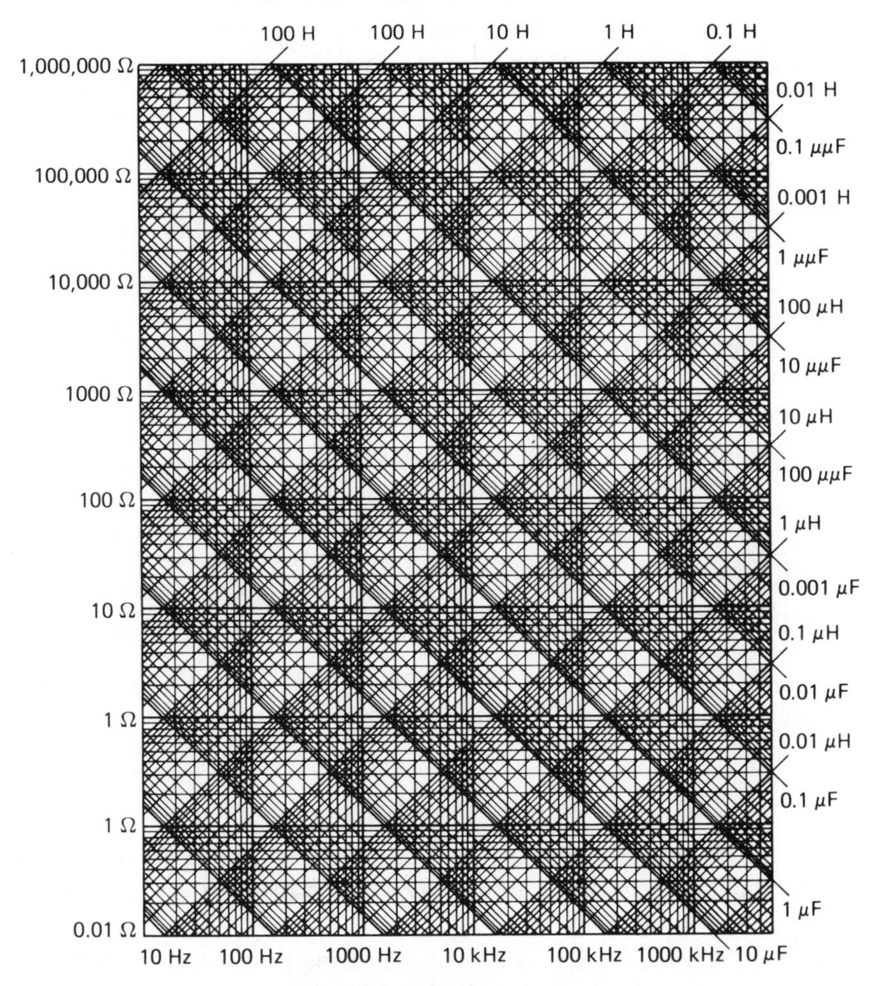

Figure B-1 Nomograph format.

most engineering requirements. Table B-4 lists temperatures from −55 °C through +155 °C. Initial resistance (R_i) is determined at 25 °C.

DEFINITIONS

Some of the more pertinent definitions that relate to terms used in magnetic-core-component design are listed in this section. Although some of the terms may seem somewhat academic to the engineer, they have been included for use by the student. Other terms are peculiar to the transformer designer and should be of interest to readers of all levels.

TABLE B-4 Resistance temperature correction factors.

To Find Resistance At:	*Multiply By:*	*To Find Resistance At:*	*Multiply By:*
−55° C	0.749	55° C	1.118
−50° C	0.761	60° C	1.138
−45° C	0.772	65° C	1.157
−40° C	0.784	70° C	1.177
−35° C	0.797	75° C	1.197
−30° C	0.810	80° C	1.216
−25° C	0.822	85° C	1.236
−20° C	0.836	90° C	1.256
−15° C	0.850	95° C	1.275
−10° C	0.865	100° C	1.295
− 5° C	0.879	105° C	1.314
0° C	0.895	110° C	1.334
5° C	0.928	115° C	1.354
10° C	0.945	120° C	1.373
15° C	0.963	125° C	1.393
20° C	0.980	130° C	1.413
25° C	1.000	135° C	1.432
30° C	1.0197	140° C	1.452
35° C	1.0393	145° C	1.471
40° C	1.590	150° C	1.491
45° C	1.0785	155° C	1.511
50° C	1.0983		

Ampère's Law: This defines the relationship that exists between current and magnetizing force. The classic equation is

$$H = \frac{0.4\pi NI}{Ml}$$

where H is the magnetizing force in oersteds, N the number of turns, I the current through the turns, and Ml the magnetic path length of the core material.

Coercive Force: A specific value of magnetizing force which is needed to lower the flux density to zero (H_c).

Faraday's Law: This defines the relationship of the voltage and flux. Thus

$$E = \frac{Nd\phi}{dt} \times 10^{-8}$$

In the case of sinusoidal voltages, it is expressed as

$$E = 2.22\phi_t FN \times 10^{-8}$$

or

$$E = 4.44 B_m A_c FN \times 10^{-8}$$

where E is the desired voltage, B_m the flux density in gausses, ϕ_t the total flux capacity of the core, A_c the effective cross-sectional area of the core, F the frequency, N the number of turns, and $\phi_t = 2B_m \times A_c$.

Gauss: This is the unit of magnetic induction used in the cgs electromagnetic system. One gauss equals 1 maxwell per cm^2.

Magnetic Flux: This is the product of the magnetic induction *(B)* and the cross-sectional area when *B* is distributed evenly and is normal to the plane of the cross section.

Maxwell: Unit of magnetic flux used in the cgs electromagnetic system. One maxwell = 10^{-8} weber.

Oersted: Unit of magnetizing force in the cgs electromagnetic system. One oersted = a magnetomotive force of 1 gilbert per cm of path length.

Permeability: Represented by the symbol μ. Broadly, this is the ratio of the changes in magnetizing force to magnetic induction (*B* to *H*).

Residual Flux: The value of magnetic induction that remains in a magnetic circuit when the magnetomotive force is lowered to zero.

Squareness Ratio: The ratio of the residual flux density to the maximum flux density (saturation).

Winding Area: Circular-mil area of the hole in a core.

Winding Factor: Ratio of the total area of wire in the center hole of a toroid to the window area of the toroid.

APPENDIX C

POT-CORE AND BOBBIN DATA

This section contains essential information for use by those who engage in the design and application of pot-core inductors and transformers. In this appendix we can find dimensional and tolerance data for a popular group of standard pot cores. The specifications are in accordance with the international standard, *IEC Publication 133,* entitled "Dimensions for Pot Cores Made of Ferromagnetic Oxides." The tables have been provided through the courtesy of Indiana General Corp.

Table C-1 contains the physical dimensions for standard pot cores. Figure C-1 provides the physical outlines for the pot cores.

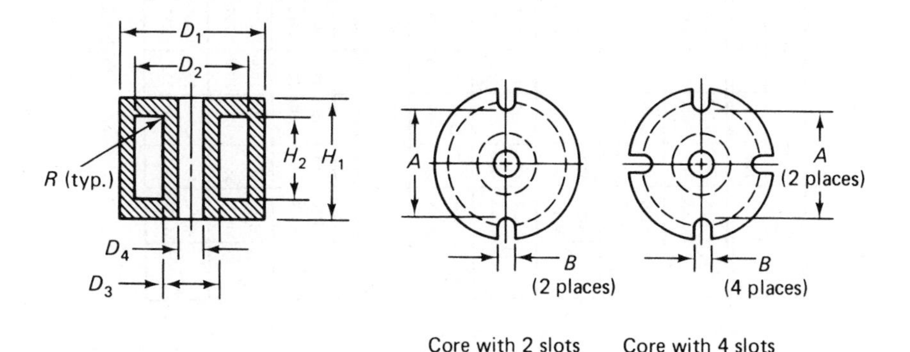

Figure C-1 Dimensional outlines for standard pot cores.

SIZE (mm)	TOL.	D_1 mm	D_1 in	D_2 mm	D_2 in	D_3 mm	D_3 in	D_4 mm	D_4 in	H_1 mm	H_1 in	H_2 mm	H_2 in	$B_{(2\ SLOT)}$ mm	$B_{(2\ SLOT)}$ in	$B_{(4\ SLOT)}$ mm	$B_{(4\ SLOT)}$ in	A mm	A in	R mm	R in
9 × 5	MIN.	9.0	.3543	.750	.2953	3.7	.1457	2.0	.0787	5.1	.2008	3.6	.1417	1.6	.063	—	—	6.0	.236	—	—
	MAX.	9.3	.3661	7.75	.3051	3.9	.1535	2.2	.0866	5.4	.2126	3.9	.1535	2.4	.094	—	—	7.2	.283	0.25	.010
11 × 7	MIN.	10.9	.429	9.0	.354	4.5	.1772	2.0	.0787	6.3	.248	4.4	.173	1.6	063	—	—	6.5	.256	—	—
	MAX.	11.3	.445	9.4	.370	4.7	.1850	2.2	.0866	6.6	.260	4.7	.185	2.6	.102	—	—	8.0	.315	0.25	.010
14 × 8	MIN.	13.8	.543	11.6	.4567	5.8	3.0	.118	8.2	.3228	5.6	.2205	2.0	.079	1.6	.063	8.7	.343	—	—	
	MAX.	14.3	.563	12.0	.4724	6.0	.2362	3.2	.126	8.5	.3346	6.0	.2362	4.1	.161	2.0	.079	10.4	.409	0.25	.010
18 × 11	MIN.	17.6	.693	14.9	.587	7.3	.2874	3.0	.118	10.4	.4094	7.2	.2835	2.0	.079	2.0	.079	11.3	.445	—	—
	MAX.	18.4	.724	15.4	.606	7.6	.2992	3.2	.126	10.7	.4213	7.6	.2992	4.4	.173	3.0	.118	14.0	.551	0.25	.010
22 × 13	MIN.	21.2	.835	17.9	.705	9.1	3583	4.4	.173	13.2	.5197	9.2	.362	2.5	098	2.5	098	13.3	.524		
	MAX.	22.0	.866	18.5	.728	9.4	.3701	4.7	.185	13.6	.5354	9.6	.378	4.4	.173	3.5	.138	16.5	.650	0.35	.014
26 × 16	MIN.	25.0	.984	21.2	.835	11.1	.437	5.4	.2126	15.9	.626	11.0	.433	2.5	.098	2.5	.098	17.0	.669	—	—
	MAX.	26.0	1.024	22.0	.866	11.5	.453	5.7	.224	16.3	.642	11.4	.449	4.4	.173	3.5	.138	20.0	.787	0.35	.014
30 × 19	MIN.	29.5	1.161	25.0	.984	13.1	.5157	5.4	.2126	18.6	.732	13.0	.5118	3.0	.118	20.0	.787	—	—		
	MAX.	30.5	1.201	25.8	1.016	13.5	.5315	5.7	.2244	19.0	.748	13.4	.5276	5.3	.209	4.0	.157	23.0	.906	0.35	.014
36 × 22	MIN.	35.0	1.378	29.9	1.177	15.6	.614	5.4	.2126	21.4	.843	14.6	.5748	3.5	.138	3.5	.138	24.0	.945	—	—
	MAX.	36.3	1.425	30.9	1.217	16.2	.638	5.7	.2244	22.0	.866	15.0	.5906	5.6	.220	4.5	.177	27.2	1.071	0.35	.014
42 × 29	MIN.	41.7	1.642	35.6	1.402	17.1	.673	5.4	.2126	29.3	1.154	20.3	.799	4.0	.160	—	—	—	—	—	—
	MAX.	43.1	1.697	37.0	1.457	17.7	.697	5.7	.2244	29.9	1.177	20.7	.815	—	—	—	—	—	—	0.40	.016

TABLE C-1 Dimensional outlines for standard pot cores.

TABLE C-2 Outline dimensions for wound bobbins.

Size (mm)	Tolerance	D_2 mm	D_2 in.	D_3 mm	D_3 in.	H_2 mm	H_2 in.	R_1 mm	R_1 in.
9 × 5	Min.	—	—	4.0	0.157	—	—	0.25	0.010
	Max.	7.4	0.291	—	—	3.6	0.142	—	—
11 × 7	Min.	—	—	4.8	0.189	—	—	0.25	0.010
	Max.	8.9	0.350	—	—	4.4	0.173	—	—
14 × 8	Min.	—	—	6.1	0.240	—	—	0.25	0.010
	Max.	11.5	0.453	—	—	5.6	0.220	—	—
18 × 11	Min.	—	—	7.7	0.303	—	—	0.25	0.010
	Max.	14.8	0.583	—	—	7.2	0.283	—	—
22 × 13	Min.	—	—	9.5	0.374	—	—	0.35	0.014
	Max.	17.8	0.701	—	—	9.2	0.362	—	—
26 × 16	Min.	—	—	11.6	0.457	—	—	0.35	0.014
	Max.	21.1	0.831	—	—	11.0	0.433	—	—
30 × 19	Min.	—	—	13.6	0.535	—	—	0.35	0.014
	Max.	24.9	0.980	—	—	13.0	0.512	—	—
36 × 22	Min.	—	—	16.3	0.642	—	—	0.35	0.014
	Max.	29.8	1.173	—	—	14.6	0.575	—	—
42 × 29	Min.	—	—	17.8	0.701	—	—	0.4	0.016
	Max.	35.5	1.398	—	—	20.3	0.799	—	—

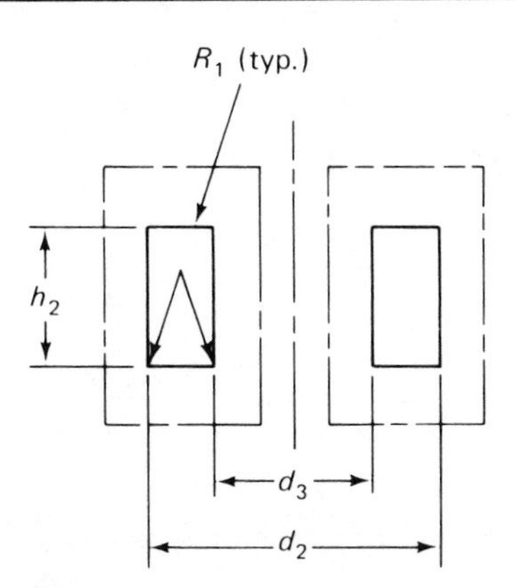

Figure C-2 Outline for a wound bobbin.

Table C-3 contains the dimensions of bobbins that will fit within the standard pot cores listed in Table C-1. Printed-circuit versions of some of these bobbins are presented in Table C-4. These dimensions are referenced to the outlines given in Fig. C-3.

See Table C-3 for complete dimensional information concerning this list of bobbins. Dimensional outlines are provided in Fig. C-4.

Table C-3 Bobbin Dimensions.

Size (mm)	No. of Sect.	Tolerance	A_1 mm²	A_1 in.²	A_2 mm²	A_2 in²	A_3 mm²	A_3 in²	B mm	B in.	D_1	D_1 in.	Example of Standard D_2 mm	D_2 in.	D_3 mm	D_3 in.	H mm	H in.
9 × 5	1	Min.	3.17	0.00492					1.6	0.060	7.23	0.285	4.67	0.184	4.01	0.158	3.40	0.134
		Max.	—	—					—	—	7.34	0.289	4.78	0.188	4.11	0.162	3.50	0.138
11 × 7	1	Min.	4.78	0.00742					1.6	0.060	8.69	0.342	5.59	0.220	4.81	0.189	4.09	0.161
		Max.	—	—					—	—	8.89	0.350	5.69	0.224	4.91	0.193	4.19	0.165
	2	Min.			2.16	0.00335												
		Max.			—	—												
	3	Min.					1.26	0.00195										
		Max.					—	—										
14 × 8	1	Min.	8.81	0.0136					1.6	0.060	11.3	0.444	6.98	0.275	5.97	0.235	5.28	0.208
		Max.	—	—					—	—	11.5	0.454	7.24	0.285	6.10	0.240	5.49	0.216
	2	Min.			3.92	0.00608												
		Max.			—	—												
	3	Min.					2.35	0.00365										
		Max.					—	—										
18 × 11	1	Min.	17.1	0.0265					1.8	0.070	14.6	0.574	8.59	0.338	7.70	0.303	6.88	0.271
		Max.	—	—					—	—	14.8	0.584	8.84	0.348	7.82	0.308	7.09	0.279
	2	Min.			7.61	0.0118												
		Max.			—	—												
	3	Min.					4.66	0.00722										
		Max.					—	—										
22 × 13	1	Min.	26.2	0.0406					1.8	0.070	17.6	0.694	10.3	0.407	9.50	0.374	8.89	0.350
		Max.	—	—					—	—	17.8	0.702	10.6	0.417	9.75	0.384	9.09	0.358
	2	Min.			12.5	0.0194												
		Max.			—	—												
	3	Min.					7.87	0.0122										
		Max.					—	—										
26 × 16	1	Min.	37.5	0.0582					1.8	0.070	20.9	0.824	12.4	0.489	11.6	0.457	10.7	0.421
		Max.	—	—					—	—	21.1	0.832	12.7	0.499	11.7	0.462	10.9	0.429
	2	Min.			17.3	0.0269												
		Max.			—	—												
	3	Min.					10.8	0.0168										
		Max.					—	—										
30 × 19	1	Min.	53.7	0.0834					1.8	0.070	24.7	0.972	14.6	0.575	13.6	0.535	12.7	0.500
		Max.	—	—					—	—	24.9	0.980	14.9	0.585	13.7	0.540	12.9	0.508
	2	Min.			25.1	0.0389												
		Max.			—	—												
	3	Min.					15.9	0.0246										
		Max.					—	—										
36 × 22	1	Min.	71.3	0.110					2.8	0.110	29.5	1.160	17.9	0.705	16.4	0.645	14.2	0.560
		Max.	—	—					—	—	29.8	1.172	18.2	0.715	16.6	0.653	14.4	0.568
	2	Min.			31.9	0.0494												
		Max.			—	—												
	3	Min.					20.0	0.0310										
		Max.					—	—										
42 × 29	1	Min.	136	0.211					2.8	0.110	35.2	1.386	19.5	0.768	18.0	0.709	19.6	0.772
		Max.	—	—					—	—	35.4	1.394	19.7	0.776	18.2	0.717	19.8	0.780
	2	Min.			55.6	0.0862												
		Max.			—	—												

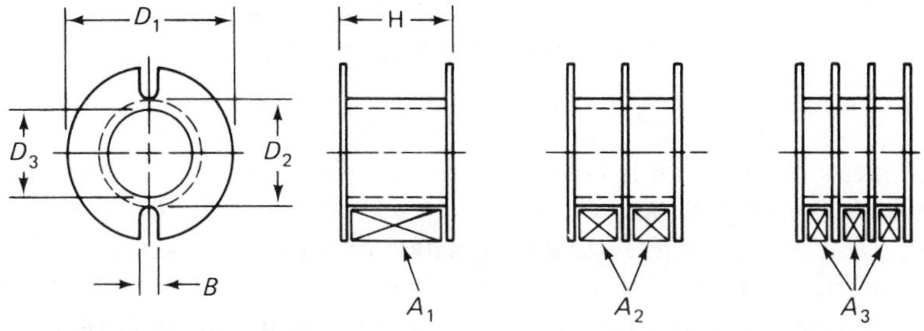

Figure C-3 Dimensional outlines for pot-core bobbins which can be used with the pot cores of Table C-1.

TABLE C-4 Printed-circuit-style bobbins.

Size (mm)	Tolerance	P mm	P in.	S_1 mm	S_1 in.	S_2 mm	S_2 in.	Figure
14 × 8	Min.	4.62	0.182	3.45	0.136	16.13	0.635	C-4C, D
	Max.	4.88	0.192	3.66	0.144	16.38	0.645	
18 × 11	Min.	4.57	0.180	3.43	0.135	21.21	0.835	C-4A, B
	Max.	4.95	0.195	3.68	0.145	21.72	0.855	
22 × 13	Min.	4.44	0.175	3.43	0.135	24.54	0.966	C-4A, B
	Max.	4.95	0.195	3.68	0.145	25.42	1.001	
26 × 16	Min.	4.62	0.182	3.45	0.136	28.07	1.105	C-4A, B
	Max.	4.82	0.192	3.71	0.146	28.88	1.137	

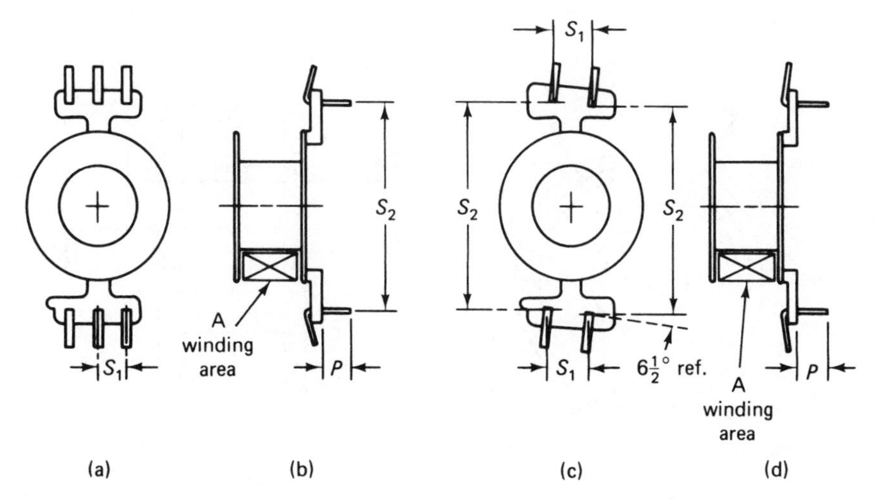

Figure C-4 Dimensional outlines for printed-circuit types of bobbins. These outlines are referenced to Table C-4.

CONVERSION INFORMATION FOR WIRE GAUGES 10 TO 44

The method used here is recommended for the calculation of the dimensional parameters of pot cores and is in accordance with *IEC Publication 205,* "Calculation of Effective Parameters of Magnetic Piece Parts."

For this method of calculating the dimensional parameters of pot cores, the pot-core set is substituted by an ideal toroidal core such that a coil wound on that toroid would give exactly the same electrical performance as a coil with same number of turns placed on the pot core set.

The dimensional parameters of that substitute toroid are called "effective" parameters. These are indicated by the suffix "*e*" added to the symbol.

Magnetic path length	l_e	mm
Cross-sectional area	A_e	mm^2
Core volume	V_e	mm^3

For the purpose of the calculation of the dimensional parameters, the closed magnetic circuit of a pot-core set is divided into five sections. For each section the area, flux path length and the core constants C_1 and C_2 are determined (Fig. C-5).

$$C_1 = \frac{l}{A}\ \mathrm{mm^{-1}} \quad \text{and} \quad C_2 = \frac{l}{A^2}\ \mathrm{mm^{-3}}$$

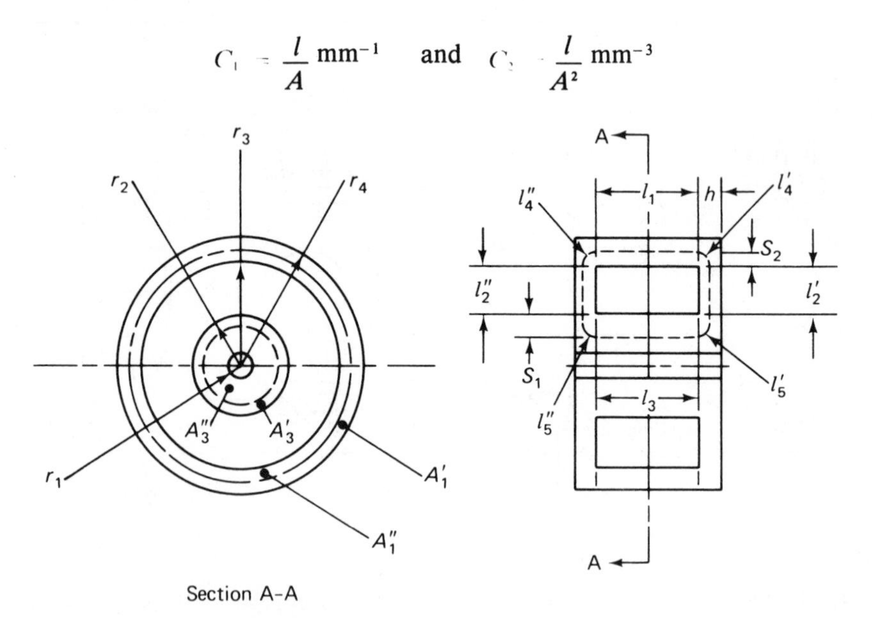

Figure C-5 Pot-core set divided into five sections.

The core constants for the total magnetic circuit of the pot-core set are

$$C_1 = \Sigma \frac{l}{A} \text{ mm}^{-1} \quad \text{and} \quad C_2 = \Sigma \frac{l}{A^2} \text{ mm}^{-3}$$

From these core constants the effective dimensional pot-core parameters can be calculated.

Magnetic path length $\ell_e = \dfrac{C_1^2}{C_2}$ mm

Cross-sectional area $A_e = \dfrac{C_1}{C_2}$ mm²

Core volume $V_e = \ell_e A_e = \dfrac{C_1^3}{C_2^2}$ mm³

For each of the five sections of the magnetic circuit of a pot-core set, the magnetic path length and cross-sectional area has to be determined:

Area of centerpost,

The condition to obtain $A'_3 = A''_3$ is

Area of outer ring,

The condition to obtain $A'_1 = A''_1$ is

Cross-sectional area of centerpost,

Cross-sectional area for outer ring,

$\dfrac{l}{A}$ for two plates, $\dfrac{l}{A^2}$ for two plates,

Mean flux path length at corners,

Cross-sectional area associated with l_4,

Mean flux path length at corners,

Cross-sectional area associated with l_5,

The calculations above ignore the effects of wire slots, which can be taken into account by the following corrections:

From A_1 subtract:

Multiply $\dfrac{l_2}{A_2}$ by

Multiply $\dfrac{l_2}{A_2^2}$ by

Multiply A_4 by

$$A_3 = A'_3 + A''_3$$

$$S_1 = r_2 - \sqrt{\frac{r_2^2 + r_1^2}{2}}$$

$$A_1 = A'_1 + A''_1$$

$$S_2 = \sqrt{\frac{r_3^2 + r_4^2}{2}} - r_3$$

$$A_3 = \pi(r_2 - r_1)(r_2 + r_1) \text{ mm}^2$$

$$A_1 = \pi(r_4 - r_3)(r_4 + r_3) \text{ mm}^2$$

$$\frac{l_2}{A_2} = \frac{1}{\pi h} \ln \frac{r_3}{r_2} = \frac{0.733}{h} \log \frac{r_3}{r_2} \text{ mm}^{-1}$$

$$\frac{\ell_2}{A_2^2} = \frac{1}{2\pi^2 h^2} \times \frac{r_3 - r_2}{r_3 r_2} \text{ mm}^{-3}$$

$$l_4 = l'_4 + l''_4 = \frac{\pi}{4}(h + 2S_2) \text{ mm}$$

$$A_4 = \frac{\pi}{2}(r_4^2 - r_3^2 + 2r_3 h) \text{ mm}^2$$

$$\ell_5 = \ell'_5 + \ell''_5 = (h + 2S_1) \text{ mm}$$

$$A_5 = \frac{\pi}{2}(r_2^2 - r_1^2 + 2r_2 h) \text{ mm}^2$$

$ng\,(r_4 - r_3)$ n = number of wire slots, g = slot width

$$\frac{1}{1 - \dfrac{ng}{2\pi r_3}}$$

$$\frac{1}{1 - \dfrac{ng}{2\pi r_3}^{2}}$$

$$1 - \frac{ng}{\pi(r_3 + r_4)}$$

APPENDIX **D**

CYLINDRICAL-CORE AND COIL DATA

The information in this appendix should be useful to those who design circuits that contain fixed-value and adjustable magnetic-core inductors. The tables offer pertinent design data and list many of the standard inductance values that are available on the commercial market. Outlines of various coil-form formats have been included to illustrate the universality of application for slug-tuned inductors.

A chart that shows recommended operating frequency versus core mixes which are available from Micrometals Corporation is shown in Fig. D-1.

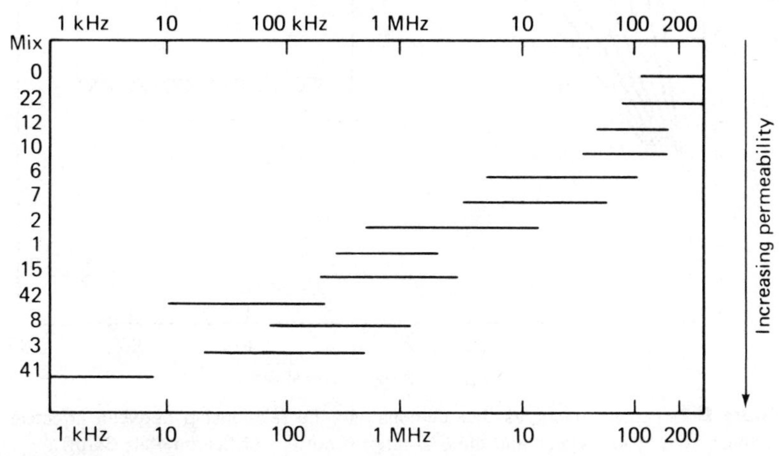

Figure D-1 Chart that shows various powdered-iron core mixes versus optimum operating frequency and *Q*.

The cores are made from powdered iron and are available in a wide range of permeabilities.

It is important to understand that the μ_e of a cylindrical core will be different from that of the same material in toroidal form. The actual μ_e is dependent upon the L/D (length-to-diameter ratio). To illustrate this effect, the curves of Fig. D-2 are included in this section. The graph is founded on a single layer winding (close-wound) occupying 95% of the cylindrical rod.

Table D-1 contains a listing of standard solid cylindrical cores that are suitable for use in slug-tuned inductors and fixed-value RF chokes. The cores are available in a wide assortment of powdered-iron mixes.

Hollow cylindrical slugs are available in a large range of powdered-iron core materials. A listing of various standard sizes is given in Table D-2.

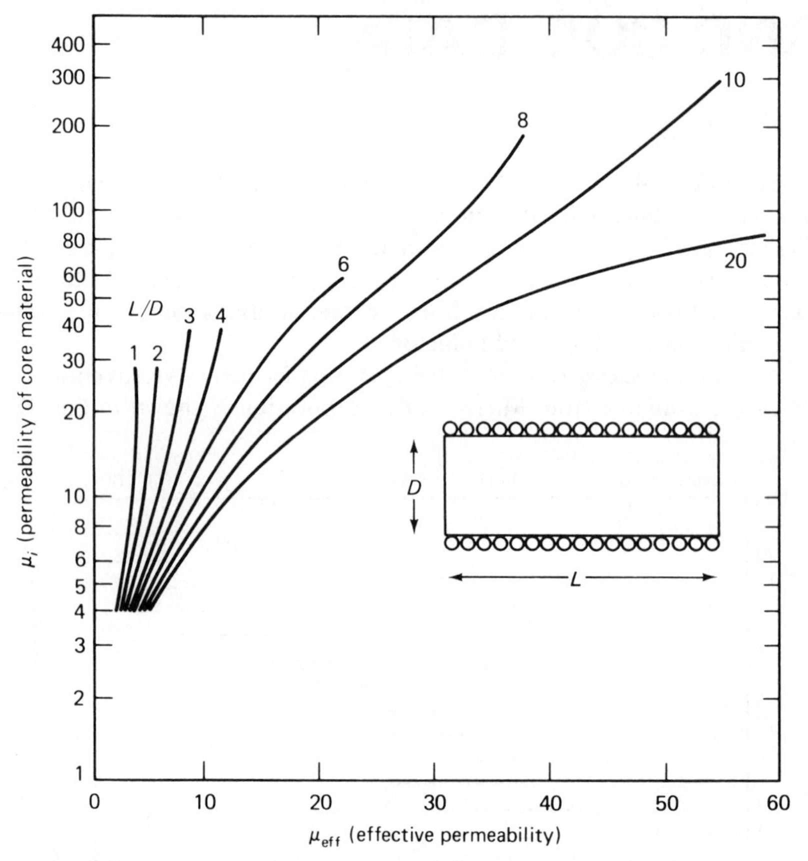

Figure D-2 Family of curves that demonstrate the relationship between effective permeability of rod material and the L/D ratio. (Courtesy of Micrometals Corp.)

TABLE D-1 Standard solid cylindrical cores.

OD (in./mm)	*Length (in./mm)*	*Part Number*
0.030/0.76	0.155/3.94	P 25-106
0.041/1.04	0.100/2.54	P 33-106
0.049/1.24	0.092/2.34	P 33-900
0.060/1.52	0.375/9.53	P 412-106
0.062/1.57	0.500/12.7	P 416-112
0.093/2.36	0.250/6.35	P 68-106
0.103/2.62	0.312/7.92	P 710-101
0.120/3.05	0.312/7.92	P 810-106
0.142/3.61	0.250/6.35	P 912-112
0.152/3.86	0.375/9.53	P 1012-102
0.177/4.50	0.625/15.8	P 1120-207
0.185/4.70	0.456/11.6	P 1214-141
0.195/4.95	0.187/4.75	P 136-207
0.245/6.22	0.250/6.35	P 168-102
0.304/7.72	1.000/25.4	P 2032-141
0.365/9.27	0.500/12.7	P 2416-103
0.370/9.40	0.750/19.0	P 2424-141
0.495/12.6	0.500/12.7	P 3216-102
0.620/15.7	1.478/37.5	P 4047-141
0.745/18.9	1.270/32.3	P 4840-102
0.995/25.2	1.000/25.4	P 6432-102

Courtesy of Micrometals Corp.

TABLE D-2 Standard hollow cylindrical slugs.

OD (in./mm)	*ID (in./mm)*	*Length (in./mm)*	*Part Number*
0.139/3.53	0.062/1.57	0.125/3.12	H 22-1006
0.257/6.53	0.075/1.91	0.375/9.52	H 46-1006
0.333/8.46	0.093/2.36	0.364/9.25	H 66-1107
0.362/9.19	0.125/3.17	0.625/15.9	H 610-1107
0.370/9.40	0.125/3.17	0.500/12.7	H 68-1006
0.370/9.40	0.205/5.21	0.500/12.7	H 68-1242
0.365/9.27	0.201/5.10	1.000/25.4	H 616-1202
0.370/9.40	0.137/3.48	1.000/25.4	H 616-1301
0.480/12.2	0.073/1.85	.500/12.7	H 88-1007
0.495/12.6	0.200/5.08	1.125/28.6	H 818-1003

Courtesy of Micrometals Corp.

Powdered-iron cylinders are also available with wire end leads embedded into them. A limited number of mixes are provided for this style of core material. A listing of the sizes is given in Table D-3. Figure D-3 shows the outline details.

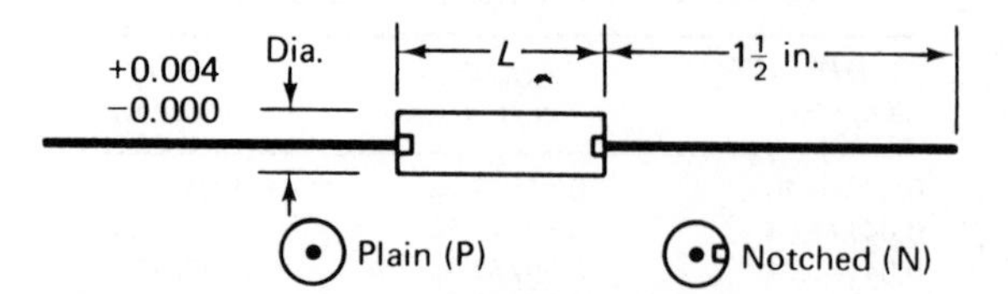

Figure D-3 Profile and end views of slug forms with wire leads. Dimensional data are given in Table D-3.

TABLE D-3 Standard powdered-iron cylinders.

OD (in./mm)	*Length (in./mm)*	*Wire Gauge*	*End D (see Fig. D-3)*	*Part Number*
0.060/1.52	0.187/4.75	24	P	F 46-20
0.060/1.52	0.250/6.35	24	P	F 48-20
0.060/1.52	0.312/7.92	24	P	F 10-20
0.069/1.75	0.203/5.16	22	P	F 57-29
0.076/1.93	0.187/4.75	22	P	F 56-20
0.076/1.93	0.250/6.35	22	P	F 58-63
0.076/1.93	0.312/7.92	22	P	F 510-23
0.076/1.93	0.375/9.53	22	P	F 512-61
0.093/2.36	0.250/6.35	22	N	F 68-21
0.093/2.36	0.312/7.92	22	N	F 610-12
0.093/2.36	0.375/9.53	22	N	F 612-21
0.103/2.62	0.250/6.35	22	N	F 78-20
0.103/2.62	0.312/7.92	22	N	F 710-14
0.103/2.62	0.375/9.53	22	N	F 711-22
0.103/2.62	0.437/11.1	22	N	F 712-24
0.121/3.07	0.312/7.92	22	N	F 810-20
0.121/3.07	0.346/8.79	22	N	F 811-70
0.121/3.07	0.375/9.53	22	N	F 812-13
0.121/3.07	0.437/11.1	22	N	F 814-20
0.121/3.07	0.500/1.27	22	N	F 816-20
0.152/3.86	0.500/12.7	21	N	F 1016-62
0.152/3.86	0.750/19.0	21	N	F 1024-40
0.186/4.72	0.500/12.7	21	N	F 1216-10
0.186/4.72	0.750/19.0	20	N	F 1224-20
0.191/4.85	0.375/9.53	22	N	F 1312-61
0.196/4.98	0.625/15.8	21	N	F 1320-20
0.216/5.49	0.750/19.0	20	N	F 1424-60
0.246/6.25	0.625/15.8	20	N	F 1620-10
0.246/6.25	0.875/22.2	20	P	F 1628-70
0.266/6.76	0.875/22.2	20	N	F 1728-20
0.280/7.11	0.875/22.2	20	N	F 1828-20
0.370/9.40	0.625/15.8	20	P	F 2420-10
0.370/9.40	1.250/31.7	16	P	F 2440-11
0.370/9.40	1.250/31.7	20	P	F 2440-12
0.495/12.6	0.500/12.7	22	N	F 3216-01

Courtesy of Micrometals Corp.

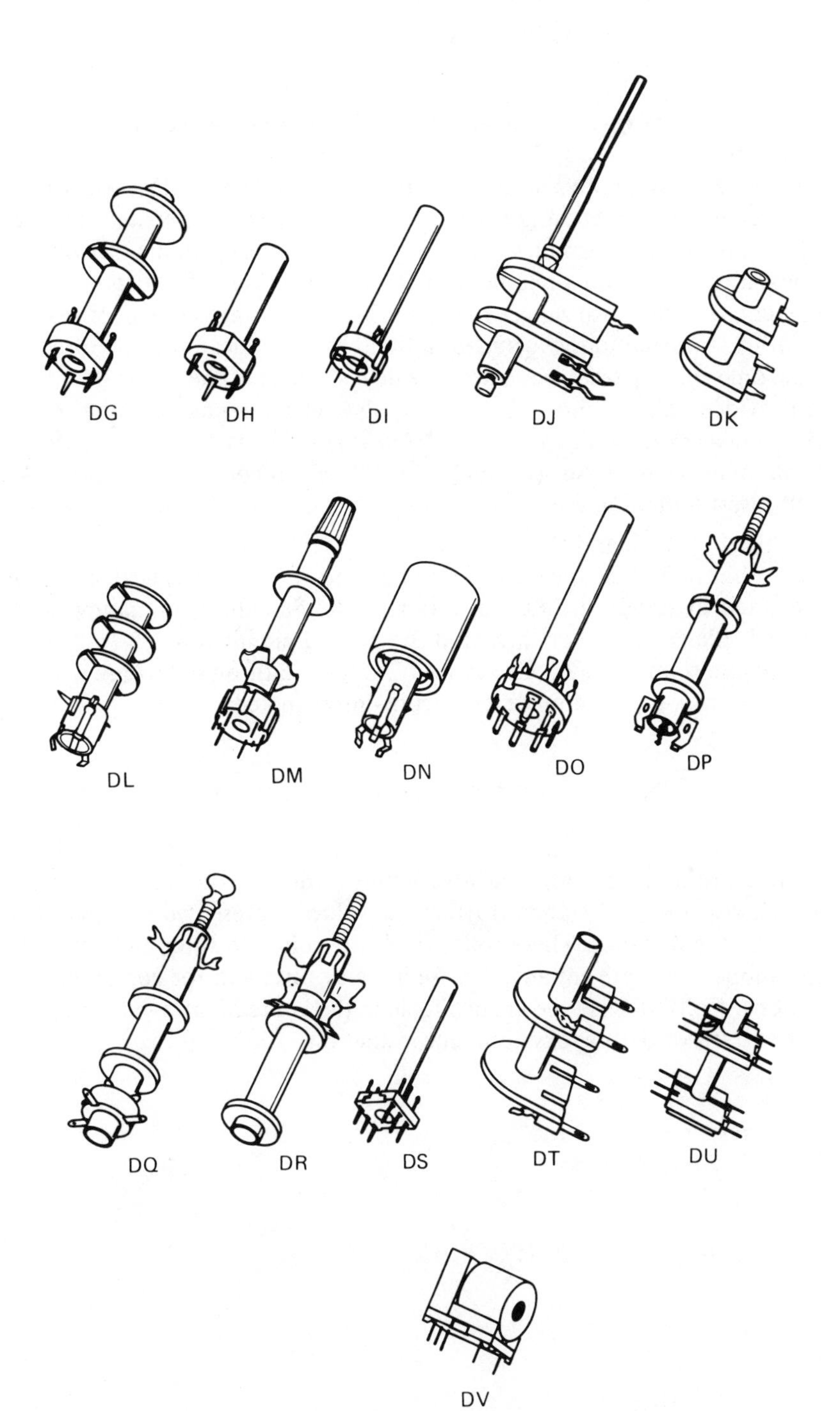

Figure D-4 (continued)

PREWOUND SLUG INDUCTORS

Standard values of variable inductance are available commercially. Printed-circuit-mount coils are sold in the United States, along with slug-tuned inductors that mount by means of a collet and nut. Most of these prefabricated units contain powdered-iron movable cores. Ferrite slugs are used in some of the assemblies to obtain large amounts of inductance (high-permeability cores) with a relatively small coil form. Table D-4 lists some of the standard inductance ranges that are offered in printed-circuit styles of coil forms. Information is included in the table to indicate the Q_u of each coil at some specific test frequency. The dc resistance of the coil windings is also provided. This group of coils uses polyester-impregnated alpha-cellulose tubing as the insulating material. Coil-form dimensions are ⅜ in. (9.52 mm) in diameter by ¾ in. (19 mm) in length. Each coil has three printed-circuit-insert mounting feet.

Tables D-5, D-6, and D-7 contain listings of three types of slug-tuned inductors that mount by means of collets and nuts. Three physical sizes are represented in standard inductance ranges: subminiature, miniature, and standard. Blank forms for these and the printed-circuit coils of Table D-4 are available commercially. The coils of Tables D-5 through D-7 are wound on slug forms which are made of silicone-impregnated ceramic.

FIXED-VALUE IRON-CORE INDUCTORS

Subminiature inductors are available with powdered-iron cores and axial leads. A wide range of standard RF choke values is presented in Table D-8 for use by the designer. These coils are ideally suited to filter and network applications. They are suitable for use in delay lines and for computer applications. Coils of this type are available in the United States with fungus-proof varnish or with epoxy-resin encapsulation.

TABLE D-4 Standard values of inductance.

Minimum Core Position			Maximum Core Position					
Minimum Inductance	*Q Min.*	*Test Frequency*	*Maximum Inductance*	*Q Min.*	*Test Frequency*	*Max. R (Ω)*	*Max. Current MHz)*	*Min. F_o (MA)*
0.095 μH	77	25.0 MHz	0.125 μH	94	25.0 MHz	0.02	4100	350.
0.130 μH	68	25.0 MHz	0.170 μH	92	25.0 MHz	0.02	1600	300.
0.185 μH	88	25.0 MHz	0.265 μH	100	25.0 MHz	0.02	1600	230.
0.285 μH	88	25.0 MHz	0.410 μH	93	25.0 MHz	0.03	1000	198.
0.420 μH	100	25.0 MHz	0.580 μH	80	25.0 MHz	0.03	2500	150.
0.540 μH	101	25.0 MHz	0.850 μH	89	25.0 MHz	0.03	1600	136.
0.640 μH	101	25.0 MHz	1.00 μH	78	25.0 MHz	0.03	1600	118.
0.760 μH	98	25.0 MHz	1.25 μH	70	7.9 MHz	0.04	1600	114.
1.20 μH	65	7.9 MHz	1.87 μH	70	7.9 MHz	0.06	1000	89.
1.65 μH	61	7.9 MHz	2.75 μH	65	7.9 MHz	0.14	400	77.
2.40 μH	64	7.9 MHz	4.10 μH	60	7.9 MHz	0.17	400	62.
3.40 μH	68	7.9 MHz	5.80 μH	60	7.9 MHz	0.24	400	53.
4.60 μH	64	7.9 MHz	8.50 μH	56	7.9 MHz	0.39	250	45
5.60 μH	64	7.9 MHz	10.0 μH	57	2.5 MHz	0.64	160	40.
7.10 μH	68	7.9 MHz	12.5 μH	55	2.5 MHz	0.77	160	38.
10.0 μH	58	2.5 MHz	18.7 μH	95	2.5 MHz	1.68	100	11.7
14.8 μH	61	2.5 MHz	27.5 μH	90	2.5 MHz	1.91	100	8.4
22.0 μH	60	2.5 MHz	41.0 μH	75	2.5 MHz	2.34	100	6.7
31.0 μH	58	2.5 MHz	58.0 μH	68	2.5 MHz	2.72	100	5.6
43.5 μH	56	2.5 MHz	85.0 μH	55	2.5 MHz	3.30	100	4.6
61.0 μH	48	2.5 MHz	100.0 μH	88	790. kHz	3.89	100	4.3
76.0 μH	52	2.5 MHz	15.0 μH	90	790. kHz	4.39	100	3.8
105.0 μH	57	790. kHz	187.0 μH	92	790. kHz	5.46	100	3.3
160.0 μH	63	790. kHz	275.0 μH	90	790. kMz	6.70	100	2.9
240.0 μH	66	790. kHz	410.0 μH	90	790. kHz	8.30	100	2.5
360.0 μH	68	790. kHz	580.0 μH	81	790. kHz	10.50	100	2.1
530.0 μH	66	790. kHz	850.0 μH	75	790. kHz	12.90	100	1.75
700.0 μH	64	790. kHz	1.00 mH	80	250. kHz	14.90	100	1.70
910.0 μH	66	790. kHz	1.25 mH	85	250. kHz	17.10	100	1.61
990.0 μH	35	250. kHz	1.87 mh	60	250. kHz	28.20	65	0.73
1.60 mH	39	250. kHz	2.75 mH	62	250. kHz	34.80	65	0.62
2.40 mH	41	250. kHz	4.10 mH	60	250. kHz	42.90	65	0.60
3.40 mH	42	250. kHz	5.80 mH	57	250. kHz	51.60	65	0.53
5.15 mH	42	250. kHz	8.50 mH	50	250. kHz	63.60	65	0.50
7.40 mH	42	250. kHz	10.00 mH	50	79. kHz	75.60	65	0.40
9.80 mH	40	250. kHz	12.50 mH	52	79. kHz	87.30	65	0.38
12.00 mH	39	79. kHz	18.70 mH	55	79. kHz	111.0	65	0.32
12.10 mH	20	79. kHz	27.50 mH	51	79. kHz	197.0	33	0.26
18.20 mH	24	79. kHz	41.00 mH	54	79. kHz	244.0	33	0.21
27.50 mH	28	79. kHz	58.00 mH	56	79. kHz	302.0	33	0.20
40.00 mH	32	79. kHz	85.00 mH	56	79. kHz	378.0	33	0.16
50.00 mH	34	79. kHz	100.00 mH	58	79. kHz	423.0	33	0.15
62.00 mH	35	79. kHz	125.00 mH	56	79. kHz	468.0	33	0.14

courtesy of J. W. Miller Co., Division of Bell Industries.

TABLE D-5 Subminiature adjustable inductors.

Minimum Core Position			Minimum Core Position					
Minimum Inductance	*Q Min.*	*Test Frequency*	*Maximum Inductance*	*Q Min.*	*Test Frequency*	*Max. R* (Ω)	*Max. Current* (*MA*)	*Min.* F_o (*MHz*)
0.190 H	64	25. MHz	0.250 H	53	25. MHz	0.014	1600	288.
0.300 H	57	25. MHz	0.390 H	44	25. MHz	0.017	1600	212.
0.440 H	61	25. MHz	0.620 H	44	25. MHz	0.031	1000	184.
0.770 H	62	25. MHz	0.900 H	41	25. MHz	0.054	636	148.
1.00 H	40	7.9 MHz	1.40 H	37	7.9 MHz	0.15	256	114.
1.60 H	40	7.9 MHz	2.40 H	36	7.9 MHz	0.25	202	92.
2.70 H	40	7.9 MHz	4.20 H	36	7.9 MHz	0.62	100	69.
4.70 H	42	7.9 MHz	6.80 H	33	7.9 MHz	0.91	100	60.
7.80 H	30	7.9 MHz	11.0 H	42	2.5 MHz	1.4	100	18.
14.0 H	34	2.5 MHz	19.0 H	37	2.5 MHz	1.9	100	13.
22.0 H	33	2.5 MHz	31.0 H	38	2.5 MHz	2.2	100	12.
36.0 H	32	2.5 MHz	49.0 H	33	2.5 MHz	3.1	100	9.6
56.0 H	23	2.5 MHz	97.0 H	24	2.5 MHz	5.6	64	7.3
111 H	26	0.79 MHz	171. H	26	0.79 MHz	7.9	64	6.0
196 H	25	0.79 MHz	285. H	24	0.79 MHz	11.	64	4.4

Courtesy of J. W. Miller Co., Division of Bell Industries.

TABLE D-6 Miniature adjustable inductors.

Minimum Core Position			Maximum Core Position					
Minimum Inductance	*Q Min.*	*Test Frequency*	*Maximum Inductance*	*Q Min.*	*Test Frequency*	*Max. R (Ω)*	*Max. Current (MA)*	*Min. F_0 (MHz)*
0.440 H	80	25. MHz	0.760 H	52	25. MHz	0.03	1600	142.
1.10 H	72	25. MHz	1.50 H	40	7.9 MHz	0.06	1000	96.
1.70 H	51	7.9 MHz	2.70 H	36	7.9 MHz	0.11	636	80.
3.10 H	56	7.9 MHz	4.80 H	33	7.9 MHz	0.23	400	58.
5.50 H	60	7.9 MHz	8.60 H	33	7.9 MHz	0.49	256	45.
9.90 H	52	7.9 MHz	15.0 H	41	2.5 MHz	1.5	100	32.
17.0 H	47	2.5 MHz	23.0 H	53	2.5 MHz	2.3	100	19.
26.0 H	48	2.5 MHz	33.0 H	51	2.5 MHz	2.9	100	16.
38.0 H	50	2.5 MHz	57.0 H	48	2.5 MHz	3.4	100	12.
66.0 H	44	2.5 MHz	114. H	40	0.79 MHz	4.1	100	5.2
120. H	46	0.79 MHz	190. H	40	0.79 MHz	5.7	100	4.1
209. H	45	0.79 MHz	314. H	32	0.79 MHz	7.7	100	3.2
350. H	53	0.79 MHz	475. H	41	0.79 MHz	10.	100	2.9
528. H	44	0.79 MHz	760. H	40	0.79 MHz	14.	100	2.5

Courtesy of J. W. Miller Co., Division of Bell Industries.

TABLE D-7 Standard adjustable inductors.

Minimum Core Position			Maximum Core Position					
Minimum Inductance	*Q Min.*	*Test Frequency*	*Maximum Inductance*	*Q Min.*	*Test Frequency*	*Max. R (Ω)*	*Max. Current (MA)*	*Min. F_0 (MHz)*
0.990 H	68	25. MHz	1.50 H	57	7.9 MHz	0.04	1600	75.
1.60 H	67	7.9 MHz	3.10 H	52	7.9 MHz	0.08	1000	51.
3.30 H	70	7.9 MHz	6.50 H	44	7.9 MHz	0.17	636	41.
7.30 H	50	7.9 MHz	14.0 H	35	2.5 MHz	0.57	256	26.
16.0 H	52	2.5 MHz	29.0 H	56	2.5 MHz	2.2	100	7.6
13.0 H	51	2.5 MHz	66.0 H	37	2.5 MHz	3.1	100	4.9
74.0 H	42	2.5 MHz	124 H	44	0.79 MHz	4.5	100	3.7
138 H	47	0.79 MHz	238 H	44	0.79 MHz	6.6	100	3.0
270 H	57	0.79 MHz	451 H	43	0.79 MHz	8.9	100	2.2
495 H	56	0.79 MHz	760 H	41	0.79 MHz	12.	100	2.0
825 H	41	0.79 MHz	1.30 mH	27	0.25 MHz	20.	100	1.7
1.40 mH	40	0.25 MH	2.00 mH	33	0.25 MHz	25.	100	1.6

Courtesy of J. W. Miller Co., Division of Bell Industries.

TABLE D-8 Iron-core inductors.

Nominal Inductance	*Minimum Q*	*Test Frequency*	*Minimum Resonant Frequency (MHz)*	*Maximum DC Resistance*	*Maximum mA Rating*	*Maximum Winding Diameter (in.)*	*Form Length + 1/32 (in.)*
0.10 μH	49	25 MHz	600	0.013	3922	0.156	5/16
0.15 μH	52	25 MHz	490	0.025	2828	0.141	5/16
0.22 μH	48	25 MHz	400	0.038	2294	0.141	5/16
0.33 μH	47	25 MHz	330	0.070	1690	0.125	5/16
0.47 μH	46	25 MHz	280	0.125	1264	0.125	5/16
0.68 μH	48	25 MHz	240	0.200	1000	0.125	5/16
0.75 μH	48	25 MHz	224	0.264	870	0.125	5/16
0.82 μH	48	25 MHz	216	0.290	830	0.125	5/16
1.0 μH	41	25 MHz	118	0.048	2041	0.165	1/4
1.2 μH	45	7.9 MHz	118	0.072	1666	0.160	1/4
1.5 μH	42	7.9 MHz	102	0.096	1443	0.160	1/4
1.8 μH	31	7.9 MHz	89	0.096	1443	0.160	1/4
2.2 μH	43	7.9 MHz	87	0.156	1132	0.160	1/4
2.7 μH	34	7.9 MHz	74	0.168	1091	0.160	1/4
3.3 μH	40	7.9 MHz	66	0.240	912	0.150	1/4
3.9 μH	35	7.9 MHz	61	0.264	870	0.150	1/4
4.7 μH	43	7.9 MHz	53	0.457	661	0.150	1/4
5.6 μH	41	7.9 MHz	49	0.492	637	0.150	1/4
6.8 μH	40	7.9 MHz	49	0.624	566	0.150	1/4
7.5 μH	32	7.9 MHz	44	0.624	566	0.150	1/4
8.2 μH	37	7.9 MHz	41	0.744	518	0.150	1/4
9.1 μH	41	7.9 MHz	21	1.44	288	0.160	1/4
10 μH	36	7.9 MHz	19	1.56	277	0.160	1/4
12 μH	52	2.5 MHz	19	1.68	267	0.160	1/4
15 μH	52	2.5 MHz	16	1.92	250	0.165	1/4
18 μH	52	2.5 MHz	15	2.28	229	0.165	1/4
22 μH	51	2.5 MHz	13	2.28	229	0.165	1/4
25 μH	48	2.5 MHz	13	2.64	213	0.170	1/4
27 μH	49	2.5 MHz	12	2.64	213	0.170	1/4
33 μH	50	2.5 MHz	10	2.76	208	0.170	1/4
39 μH	48	2.5 MHz	9.3	3.36	188	0.175	1/4
47 μH	44	2.5 MHz	9.1	3.36	188	0.175	1/4
56 μH	45	2.5 MHz	8.6	3.84	176	0.180	1/4
68 μH	42	2.5 MHz	8.1	4.20	169	0.180	1/4
75 μH	38	2.5 MHz	7.2	4.56	162	0.185	1/4
82 μH	41	2.5 MHz	6.7	4.80	158	0.185	1/4
82 μH	41	2.5 MHz	6.7	4.92	156	0.185	1/4
100 μH	25	2.5 MHz	3.6	7.68	139	0.165	1/4
120 μH	40	790 kHz	3.2	8.16	135	0.165	1/4
150 μH	47	790 kHz	3.0	8.16	135	0.165	1/4

TABLE D-8 (continued)

Nominal Inductance	*Minimum Q*	*Test Frequency*	*Minimum Resonant Frequency (MHz)*	*Maximum DC Resistance*	*Maximum mA Rating*	*Maximum Winding Diameter (in.)*	*Form Length + 1/32 (in.)*
180 μH	48	790 kHz	2.8	8.16	135	0.170	1/4
200 μH	47	790 kHz	2.7	10.3	120	0.170	1/4
220 μH	46	790 kHz	2.5	11.5	114	0.170	1/4
250 μH	49	790 kHz	2.5	12.1	111	0.170	1/4
270 μH	46	790 kHz	2.5	13.2	106	0.175	1/4
300 μH	46	790 kHz	2.2	13.2	106	0.175	1/4
330 μH	41	790 kHz	2.0	13.9	103	0.175	1/4
350 μH	46	790 kHz	2.0	14.4	102	0.180	1/4
390 μH	45	790 kHz	2.0	15.8	97	0.180	1/4
470 μH	35	790 kHz	1.8	16.3	95	0.185	1/4
μ							
500 μH	49	790 kHz	1.8	18.0	91	0.195	1/4
560 μH	41	790 kHz	1.7	19.2	88	0.195	1/4
680 μH	37	790 kHz	1.6	19.8	87	0.200	1/4
750 μH	40	790 kHz	1.6	22.9	80	0.210	1/4
820 μH	33	790 kHz	1.6	22.9	80	0.210	1/4
910 μH	32	790 kHz	1.4	24.0	79	0.220	1/4
1.00 mH	30	790 kHz	1.4	24.0	79	0.225	1/4
1.20 mH	34	250 kHz	1.2	33.6	66	0.220	1/4
1.50 mH	40	250 kHz	1.1	37.2	63	0.225	1/4
1.80 mH	40	250 kHz	0.96	42.0	59	0.235	1/4
2.20 mH	40	250 kHz	0.96	45.6	57	0.240	1/4
2.50 mH	48	250 kHz	0.96	45.6	57	0.260	3/8
2.70 mH	50	250 kHz	0.88	45.6	57	0.260	3/8
3.30 mH	52	250 kHz	0.80	51.6	53	0.260	3/8
3.90 mH	53	250 kHz	0.76	57.6	51	0.275	2/8
4.70 mH	49	250 kHz	0.68	64.8	48	0.285	3/8
5.60 mH	53	250 kHz	0.68	69.6	46	0.300	3/8
6.80 mH	51	250 kHz	0.64	78.0	43	0.310	3/8
7.50 mH	49	250 kHz	0.60	85.2	41	0.310	3/8
8.20 mH	48	250 kHz	0.60	92.4	40	0.330	3/8
9.10 mH	52	250 kHz	0.56	98.4	39	0.330	3/8
10.0 mH	41	250 kHz	0.52	101	38	0.335	3/8
12 mH	46	79 kHz	0.36	100	50	0.300	1/2
15 mH	50	79 kHz	0.32	113	47	0.300	1/2
18 mH	49	79 kHz	0.29	128	44	0.325	1/2
22 mH	50	79 kHz	0.27	144	41	0.330	1/2
25 mH	59	79 kHz	0.250	115	46	0.340	5/8
27 mH	61	79 kHz	0.244	120	45	0.353	5/8
33 mH	61	79 kHz	0.232	134	43	0.353	5/8
39 mH	59	79 kHz	0.220	147	41	0.370	5/8
47 mH	57	79 kHz	0.206	168	38	0.384	5/8
50 mH	57	79 kHz	0.196	175	37	0.400	5/8
56 mH	57	79 kHz	0.188	189	36	0.400	5/8
68 mH	57	79 kHz	0.180	215	34	0.415	5/8
75 mH	53	79 kHz	0.174	222	33	0.430	5/8
82 mH	50	79 kHz	0.168	238	32	0.430	5/8
91 mH	51	79 kHz	0.166	250	31	0.430	5/8
100 mH	48	79 kHz	0.157	278	29	0.446	5/8

Courtesy of J. W. Miller Co., Division of Bell Industries.

APPENDIX E

TOROID-CORE DATA

This material is provided as a supplement to the information on toroids that was presented earlier in the book. These data will be useful to engineers and students who undertake the design of toroidal inductors and transformers. Various core sizes and types are specified in this appendix, along with pertinent operating curves.

Table 1-1 provides data on optimum frequencies and color codes for powdered-iron materials which have μ_i characteristics from 1 to 75. Table 1-7 lists similar information for ferrite toroids.

PERMEABILITY VERSUS DC BIAS

Figure E-1 contains a set of curves for powdered-iron toroids which have permeabilities ranging from 10 to 100. These curves demonstrate the *H* characteristics of the cores versus other pertinent parameters.

PERMEABILITY VERSUS AC FLUX DENSITY

Figure E-2 presents a family of curves that relate to factor *B* in toroid cores. The initial permeability is rendered in percentage versus the ac flux density. These data are useful for quick reference in a design exercise. It illustrates clearly the effect flux density has on the core permeability.

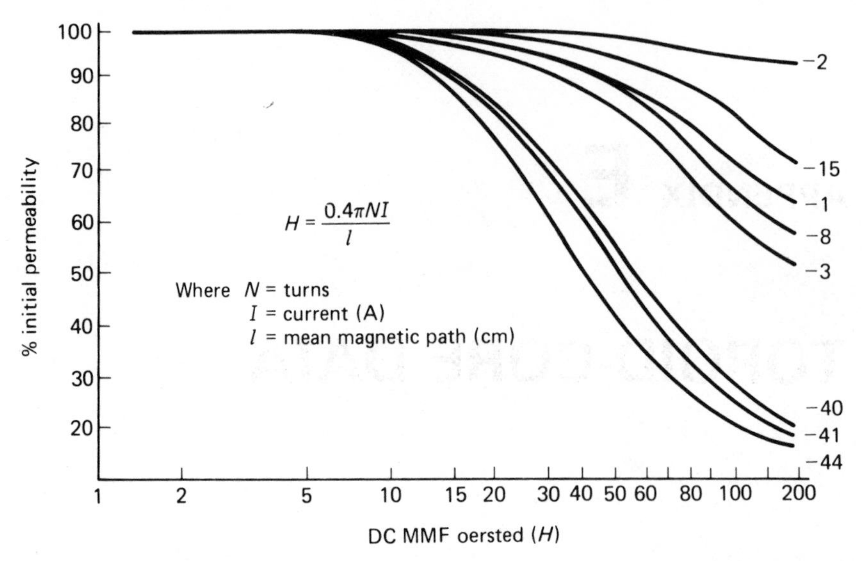

Figure E-1 Permeability versus dc bias for various powdered-iron core materials. (Courtesy of Micrometals Corp.)

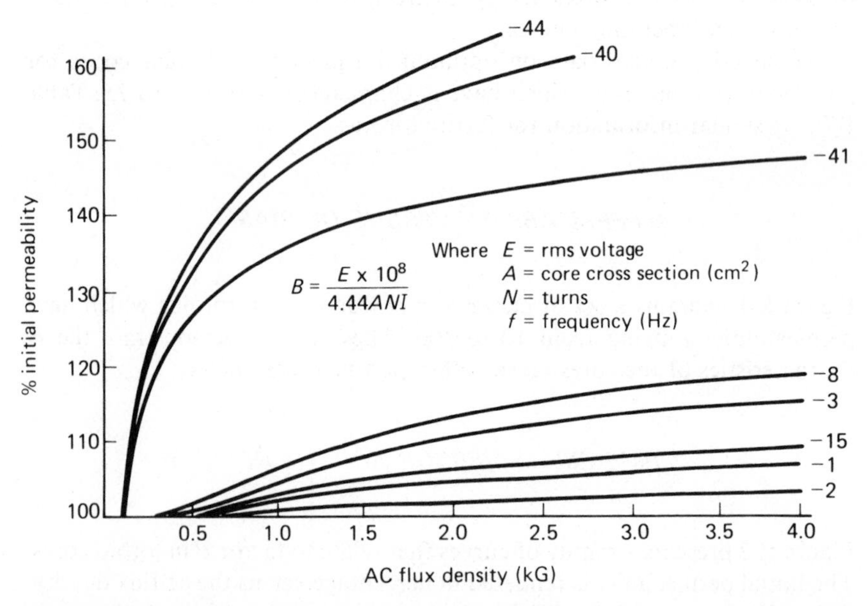

Figure E-2 Percentage of initial permeability versus ac flux density for powdered-iron toroids. (Courtesy of Micrometals Corp.)

TEMPERATURE CHARACTERISTICS

It was established in the early chapters of this book that the core temperature of a transformer or inductor has a significant effect on the permeability, which in turn affects the inductance of the core windings. Figures E-3 through E-6 contain curves for the main group of powdered-iron toroid cores manufactured by Micrometals Corp. The curves show the various core mixes versus temperature and percentage of inductance change.

TOROID-CORE FILL FACTOR

It is necessary for the designer to know the fill factor of a given toroid core in order to make a proper core selection. This section explains how this is determined and provides data on the various wire gauges versus fill factor.

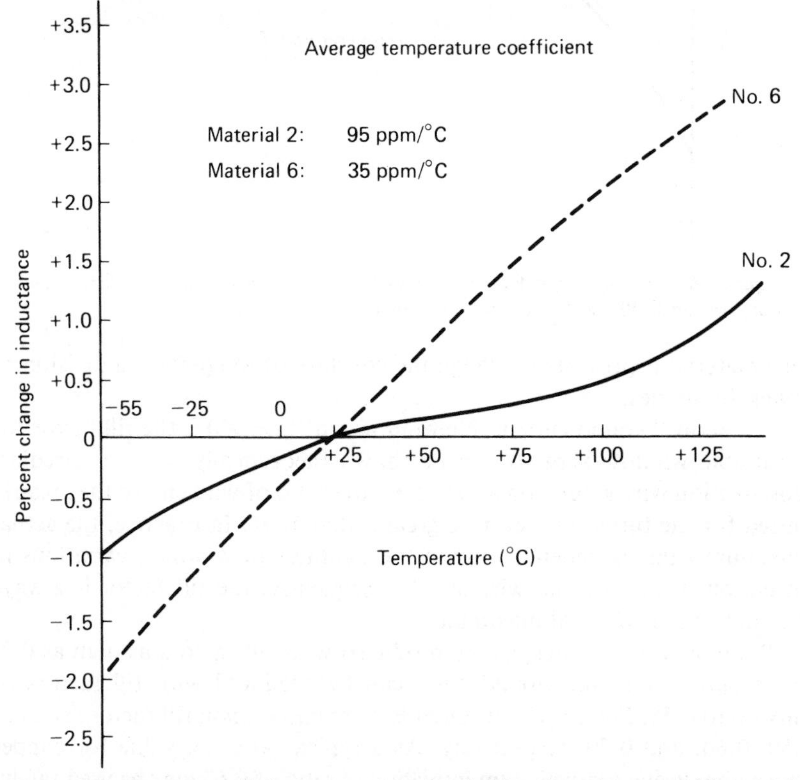

Figure E-3 Temperature/inductance curves for Amidon Associates and Micrometals Corp. core types 2 and 6. (Courtesy of Amidon Associates.)

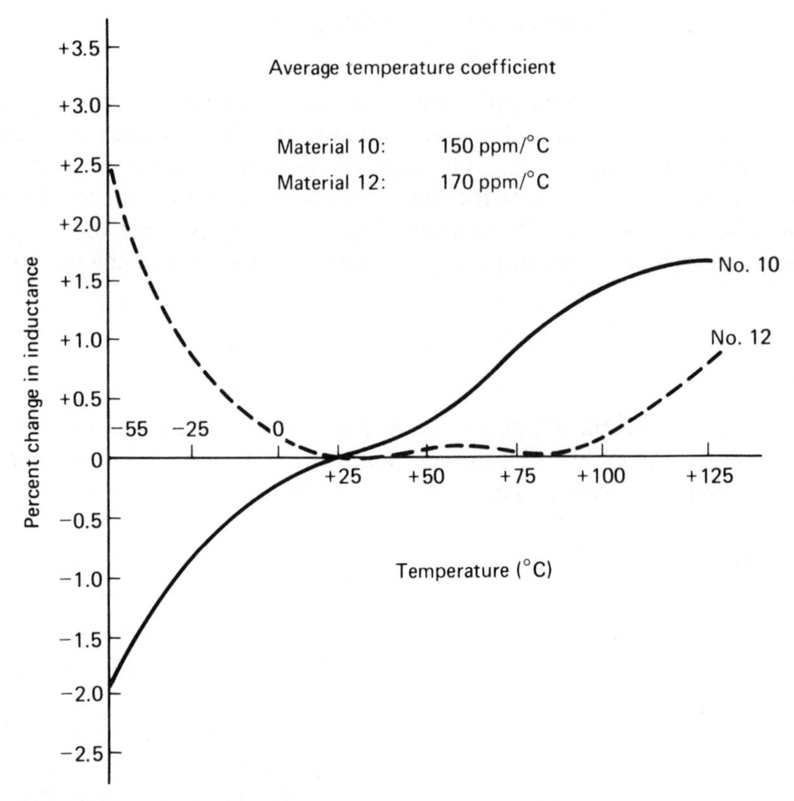

Figure E-4 Temperature/inductance curves for Amidon Associates and Micrometals Corp. materials 10 and 12. (Courtesy of Amidon Associates.)

This material is presented through the courtesy of Magnetics, a division of Spang Industries.

The term "wound cm²/usable window cm²" (= K_2) is the fill factor for the usable window area. It can be shown theoretically that for circular-cross-section wire wound on a flat form, the ratio of wire cm² to the area required for the turns can never be greater than 0.91. In practice, the actual maximum value is dependent upon the tightness of winding, variations in insulation thickness, and wire lay. Consequently, the fill factor is always less than the theoretical maximum.

Random-wound cores can be produced with fill factors as high as 0.7, but progressive sector wound cores can be produced with fill factors of only up to 0.55. Figures E-7 through E-9 are based upon fill factor ratios of 0.50, 0.60, and 0.70, respectively. As a typical working value for copper wire with a heavy synthetic film insulation, a ratio of 0.60 may be used safely.

The term "usable window cm²/window cm²" (= K_3) defines how much

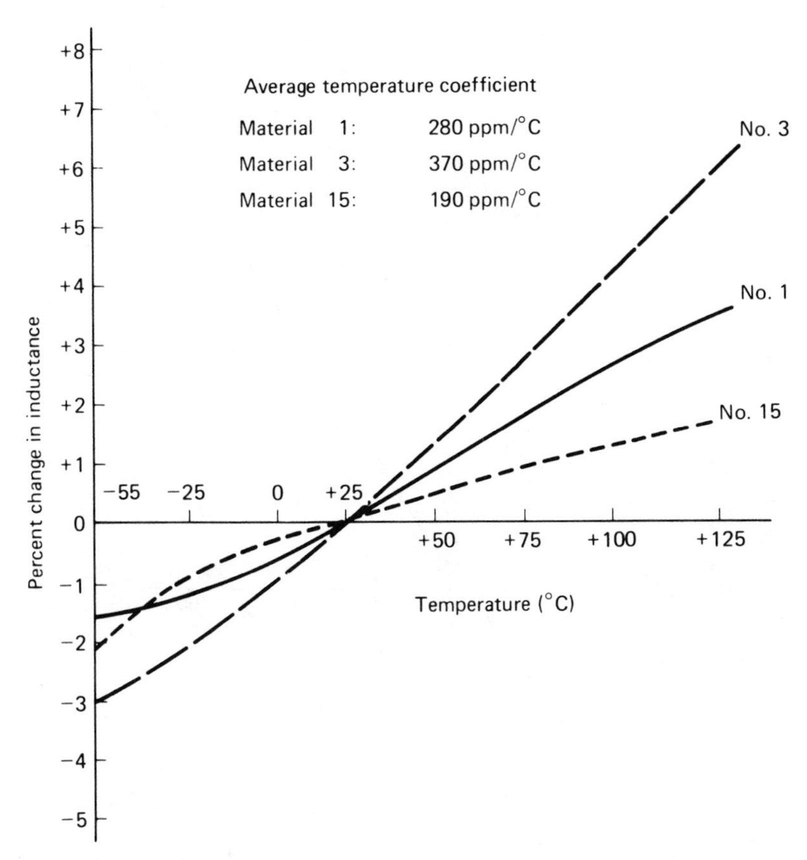

Figure E-5 Temperature/inductance curves for Amidon Associates and Micrometals Corp. core types 1, 3, and 15. (Courtesy of Amidon Associates.)

of the available window space may actually be used for the winding. The charts are based on the assumption that the inside diameter of the wound core is one-half that of the bare core; that is, $K_3 = 0.75$ (to allow free passage of the shuttle) (Fig. E-10). A typical value for the copper fraction in the window area is about 0.40. For example, for AWG 20 wire, $K_1 \times K_2 \times K_3 = 0.855 \times 0.60 \times 0.75 = 0.385$.

CORE SIZE SELECTION

Upon selection of the transformer core material and material thickness, the next step is to select the proper size core for a transformer with a given operating frequency and output power. The power-handling capability of a

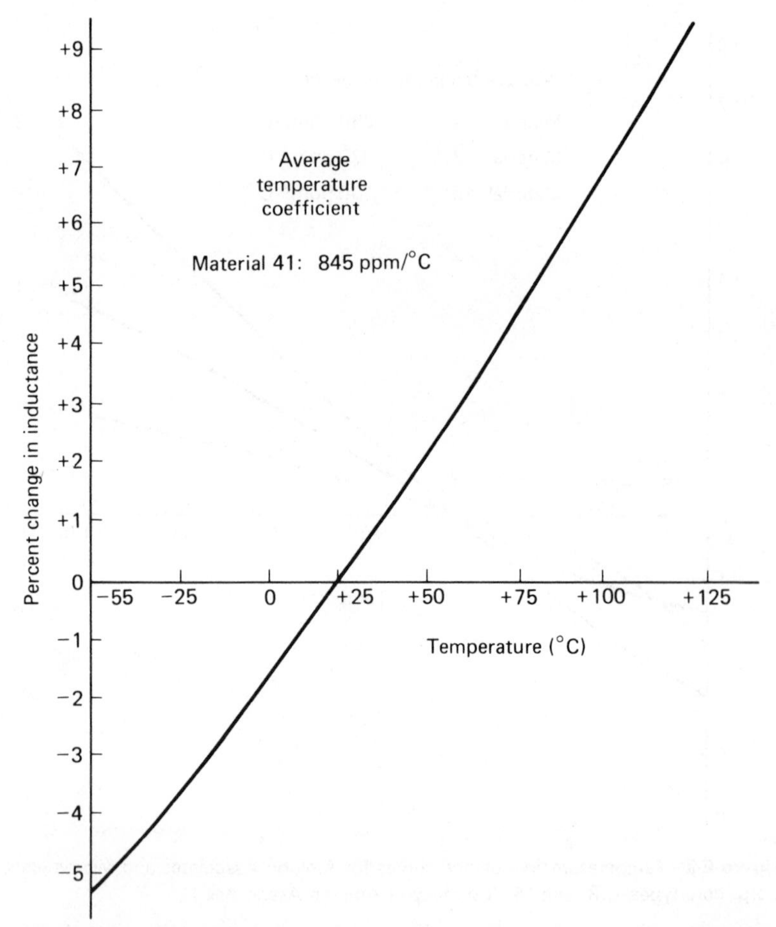

Figure E-6 Temperature/inductance curves for Amidon Associates and Fair-Rite Corp. ferrite-core material 41. (Courtesy of Amidon Associates.)

transformer can be determined by its W_aA_c product, where W_a is the available core window area in cm², and A_c the core effective cross-sectional area in cm².

The W_aA_c relationships are obtained by solving Faraday's law in the following manner:

$$\text{Faraday's law} = E = 4B_mA_cNf \times 10^{-4} \text{ square wave}$$

$$E = 4.4B_{mm}A_cNf \times 10^{-4} \text{ sine wave}$$

where E = applied voltage (rms)

B_m = flux density in teslas

A_c = core effective cross-sectional area in cm²

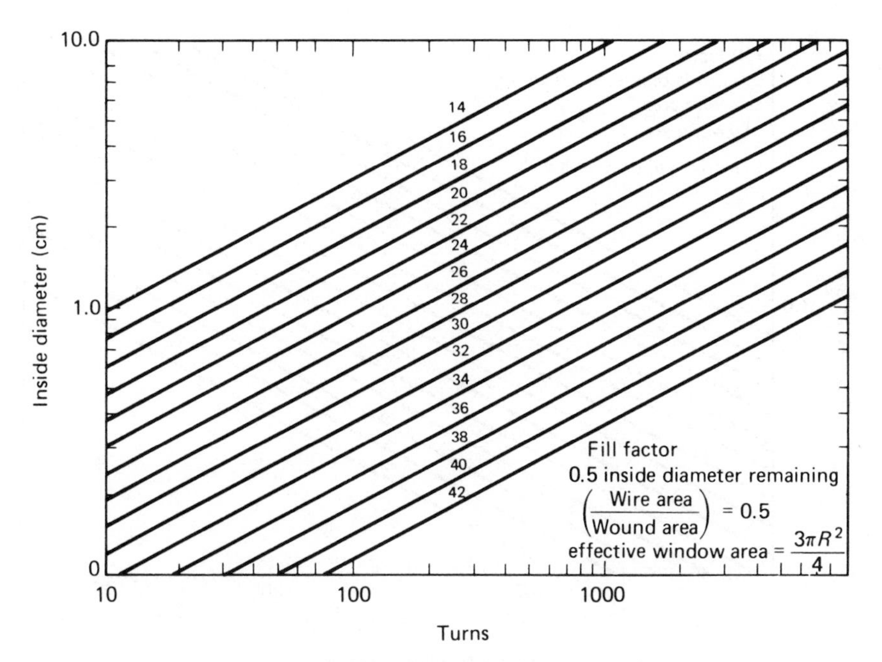

Figure E-7 Fill factor 0.5.

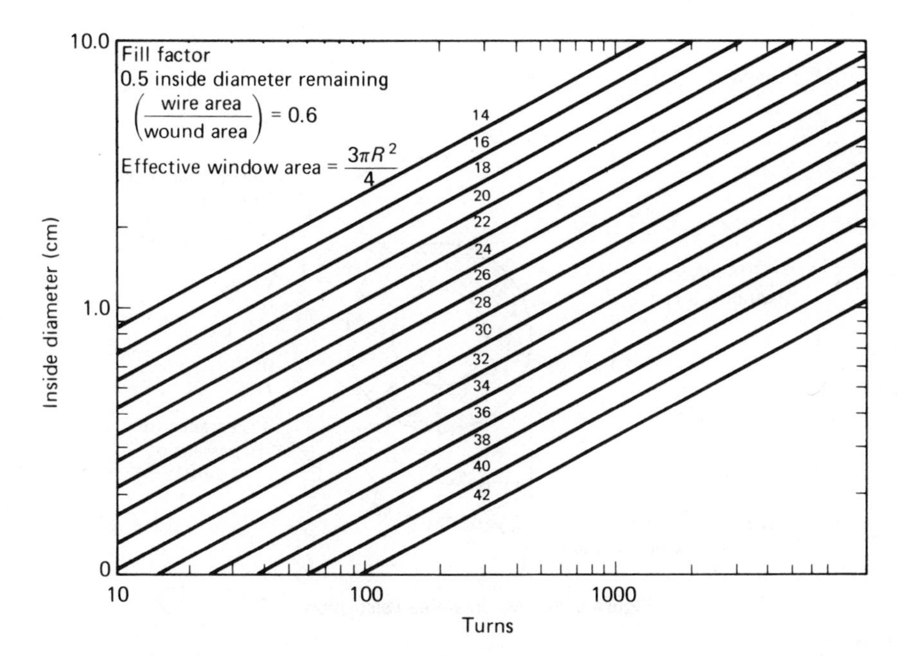

Figure E-8 Fill factor 0.6.

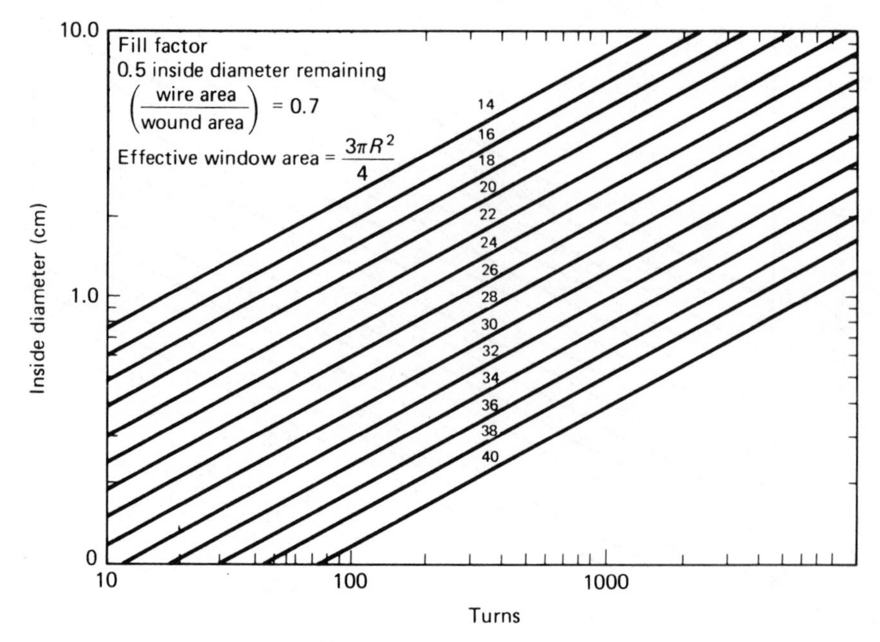

Figure E-9 Fill factor 0.7.

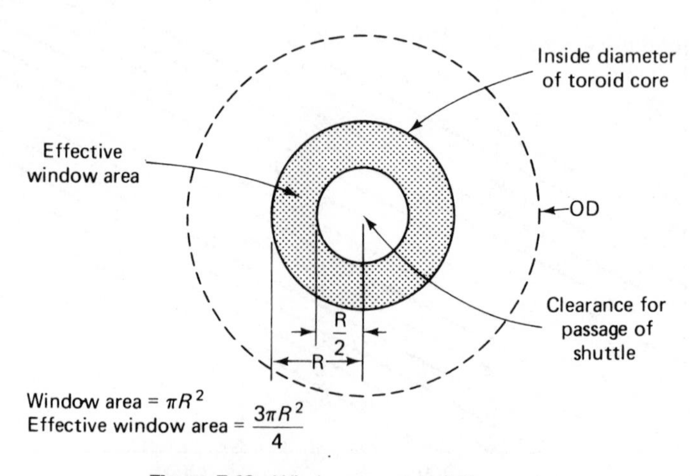

Figure E-10 Window-area calculation.

N = number of turns
f = frequency in hertz
A_w = bare wire area in cm^2
W_a = window area in cm^2
K = window utilization factor
I = current (rms)
J = current density
P_0 = output power (total)
P_1 = input power
P_T = total power
η = efficiency

Solving yields

$$NA_c = \frac{E \times 10^4}{4B_m f}$$

Window utilization factor:

$$K = \frac{NA_w}{W_a}$$

$$N = \frac{KW_a}{A_w}$$

Multiply both sides by A_c:

$$NA_c = \frac{KW_aA_c}{A_w} = \frac{E \times 10^4}{4B_m f}$$

Combining and solving for $W_a\ A_c$ yields

$$\frac{KW_aA_c}{A_w} = \frac{E \times 10^4}{4B_m fK}$$

$$W_aA_c = \frac{EA_w \times 10^4}{4B_m fK}$$

$$J = \frac{I}{A_w} = \frac{\text{A}}{\text{cm}^2}$$

$$\eta = \frac{P_0}{P_1}$$

$$P_1 = EI$$

$$EA_w = \frac{EI}{J} = \frac{P_1}{J} = \frac{P_0}{J\eta}$$

$$\underset{(total)}{W_aA_c} = \underset{(primary)}{W_aA_c} + \underset{(secondary)}{W_aA_c}$$

$$\underset{(total)}{W_aA_c} = \frac{P_0}{J} \times \frac{10^4}{4B_m fK} + \frac{P_0 \times 10^4}{4B_m fKJ} = \frac{P_0 \times 10^4}{4B_m fKJ}(1/\eta + 1)$$

$$P_T = \frac{P_0}{\eta} + P_0$$

$$W_aA_c = \frac{P_T \times 10^4}{4B_m fKJ}$$

Window utilization factor K:

Lamination and bobbin

toroid 1/2 I.D. remaining $\frac{W_a\ (eff.)}{W_a} \times \text{fill factor} \times \frac{A_w \text{ bare}}{A_w \text{ total}} = 0.4$
C core and bobbin

$$W_aA_c = \frac{P_T \times 10^4}{1.6 \times B_m fJ} \quad \text{square wave}$$

$$W_aA_c = \frac{P_T \times 10^4}{1.77 \times B_m fJ} \quad \text{sine wave}$$

The curve in Fig. E-11 shows the required core W_aA_c product plotted against transformer output power for different frequency. The values held constant were

$$Bm = 0.3T$$
$$J = 200 \text{ A/cm}^2$$
$$K = 0.40$$
$$\eta = 95\%$$

These values were held constant so that one nomograph could be used with all materials and the engineer could adjust B_m, J, K, and η to fit the design. From the equation

$$W_aA_c = \frac{P_T \times 10^4}{1.6 \times B_m fJ}$$

two nomographs were generated to compare output power

$$P_0 = \frac{P_T}{1/\eta + 1}$$

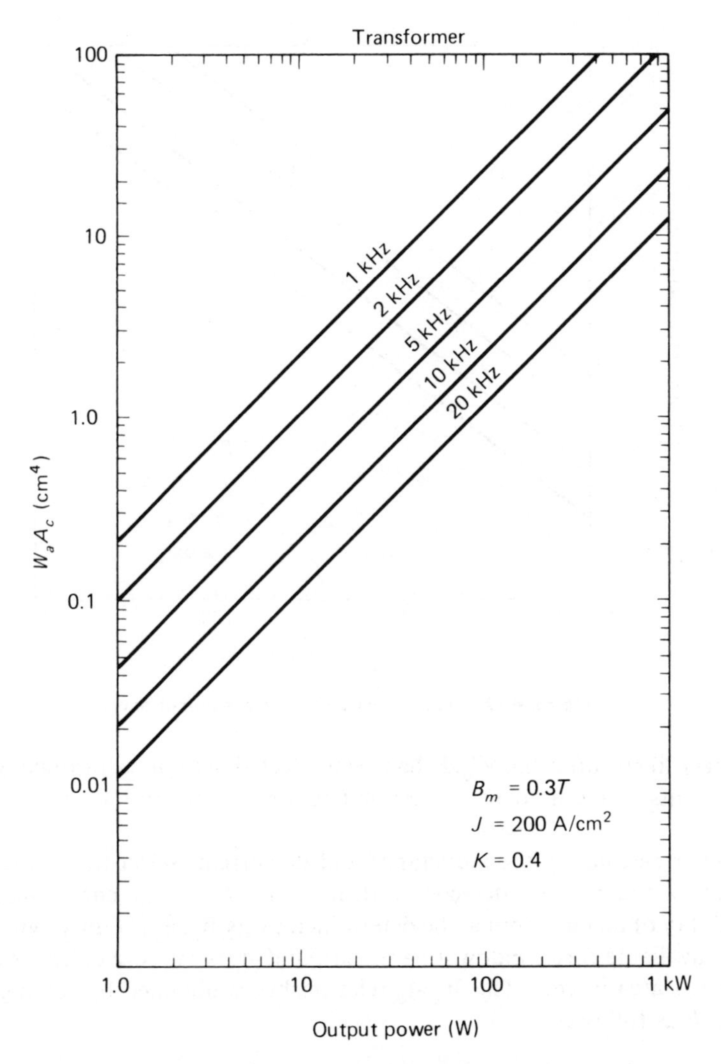

Figure E-11 W_aA_c versus output power.

with W_aA_c and P_0 with weight. These nomographs were generated from the lamination and C core in the article. The nomograph in Fig. E-11 compares P_0 with W_aA_c; thus the size of the transformer can quickly be determined. The nomograph in Fig. E-12 compares power with weight; thus the weight of a fully wound transformer can quickly be determined. These nomographs have the following constraints:

$$B_m = 0.3T$$
$$J = 200 \text{ A/cm}^2$$
$$K = 0.4$$

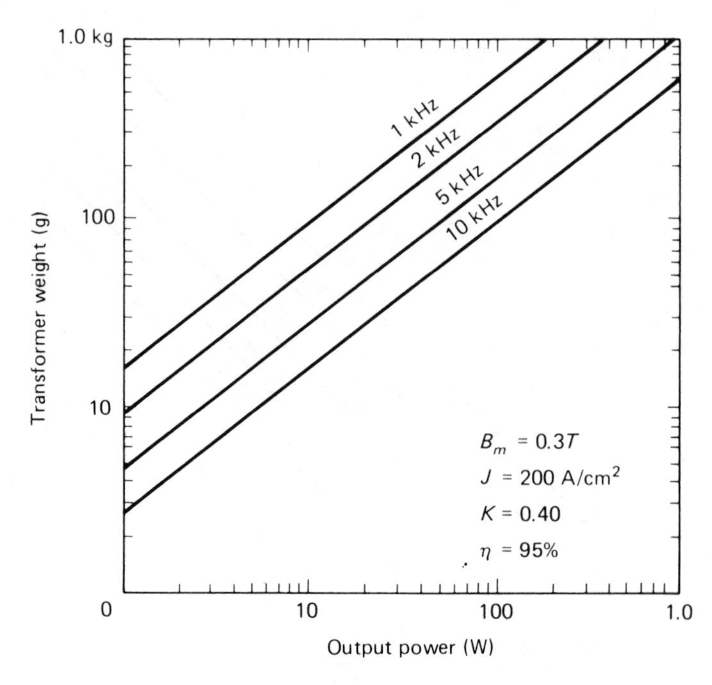

Figure E-12 Transformer weight versus output power.

Very likely, after the W_aA_c has been selected, a slight adjustment of the constraints is required to correspond to the actual core W_aA_c products available.

After calculating the inductance and dc current, select the proper permeability and size powder core with a given $LI^2/2$. The energy-handling capability of an inductor can be determined by its W_aA_c product, where W_a is the available core window area in cm², and A_c is the core effective cross-sectional area in cm². The W_aA_c relationship is obtained by solving $E = L\,dI/dt$ as follows:

where E = voltage, volts
L = inductance, henries
I = current, amperes
N = number of turns
ϕ = flux, webers
B_m = flux density, teslas
$A'_c \doteq$ core cross section, m²
μ_r = relative permeability
μ_o = absolute permeability
$(4\pi \times 10^{-7})$

H' = magnetizing force, amp turns/m
ℓ'_m = magnetic path length, m
K = window utilization factor
W'_a = window area, m²
J' = current density, A/m²
Eng = energy, watt seconds

$$E = L\frac{dI}{dt} = N\frac{d\phi}{dt}$$

$$L = N\frac{d\phi}{dI}$$

$$\phi = B_m A'_c$$

$$B_m = \mu_r\mu_0 H' = \frac{\mu_r\mu_0 NI}{\ell'_m}$$

$$\phi = \frac{\mu_r\mu_0 NIA'_c}{\ell'_m}$$

$$\frac{d\phi}{dI} = \frac{\mu_r\mu_0 NA'_c}{\ell'_m}$$

$$L = N\frac{d\phi}{dI} = \frac{\mu_r\mu_0 N^2 A'_c}{\ell'_m}$$

$$\text{Energy} = \tfrac{1}{2} LI^2 = \frac{\mu_r\mu_0 N^2 A'_c}{2\ell'_m} I^2$$

If B_m is specified,

$$I = \frac{B_m\ell'_m}{\mu_r\mu_0 N}$$

$$\text{Eng} = \frac{\mu_r\mu_0 N^2 A'_c}{2\ell'_m}\ \frac{B_m\ell'_m}{\mu_r\mu_0 N}^2 = \frac{(B_m)^2\ell'_m A'_c}{2\mu_r\mu_0} \qquad \text{watt seconds}$$

$$I\frac{KW'_aJ'}{N} = \frac{B_m\ell'_m}{\mu_r\mu_0 N}$$

Solving for $\mu_r\mu_0$ yields

$$\mu_r\mu_0 = \frac{B_m\ell'_m}{KW'_aJ'}$$

Substituting into the energy equation, we obtain

$$\text{Eng} = \frac{(B_m)^2\ \ell'_m A'_c}{2} \times \frac{KW'_aJ'}{B_m\ell'_m} = \frac{W'_aA'_cB_mJ'K}{2}$$

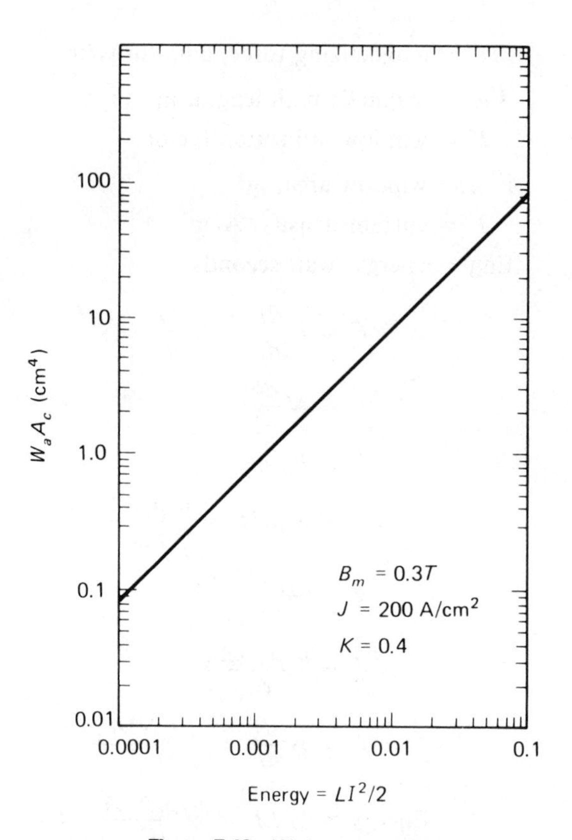

Figure E-13 W_aA_c versus $LI^2/2$.

Let

$$W_a = \text{window area, cm}^2$$
$$A_c = \text{core area, cm}^2$$
$$J = \text{current density, A/cm}^2$$
$$W'_a = W_a \times 10^{-4}$$
$$A'_c = A_c \times 10^{-4}$$
$$J' = J \times 10^4$$

Substituting into the energy equation yields

$$\text{Eng} = \frac{W_aA_cB_mJK}{2} \times 10^{-4}$$

Solving for W_aA_c, we obtain

$$W_aA_c = \frac{2\,(\text{Eng})}{B_mJK} \times 10^4$$

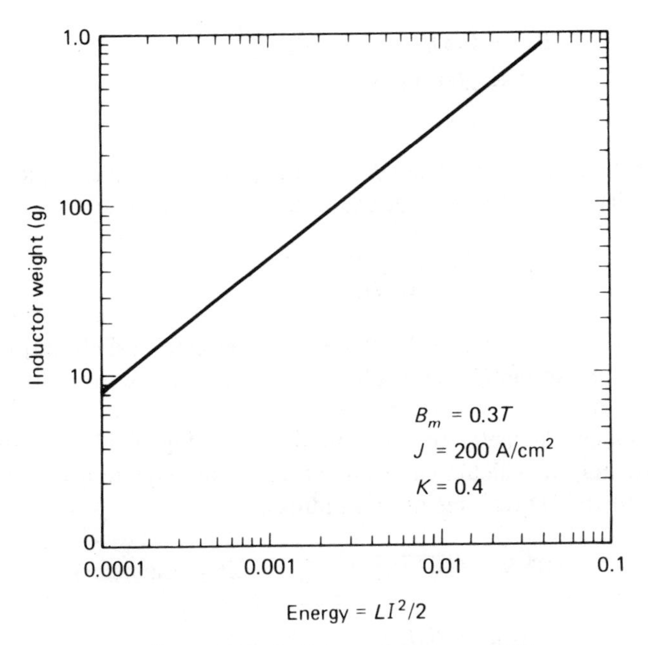

Figure E-14 Inductor weight versus $LI^2/2$.

Let

$$\ell_m = \text{magnetic path length, cm}$$

$$\ell'_m = \ell_m \times 10^{-2}$$

$$\mu_r = \frac{B_m \ell_m \times 10^{-2}}{K\mu_0(W_a \times 10^{-4})(J \times 10^4)} = \frac{B_m \ell_m \times 10^{-2}}{\mu_0 \mathrm{W_a JK}}$$

For $\mu_0 = 4\pi \times 10^{-7}$,

$$\mu_r = \frac{B_m \ell_m \times 10^{-2}}{4\pi \times 10^{-7}\, W_a JK} = \frac{B_m \ell_m \times 10^4}{0.4 W_a JK}$$

From the equation

$$W_a A_c = \frac{2(\text{Eng})}{B_m JK}$$

two nomographs were generated to compare energy or $LI^2/2$ with $W_a A_c$ and energy or $LI^2/2$ with weight. These nomographs were generated for 13 commonly used powder cores in this article. The nomograph in Fig. E-13 compares $LI^2/2$ with $W_a A_c$; thus the size of an inductor can quickly be determined. The nomograph in Fig. E-14 compares $LI^2/2$ with weight; thus the weight of a fully wound inductor can quickly be determined. These nomographs have the following constraints:

$$B_m = 0.3T$$
$$J = 200 \text{ A/cm}^2$$
$$K = 0.4$$

After the core size has been determined, the next step is to pick the right permeability for that core size, using the following equation:

$$\mu_r = \frac{B_m \ell_m \times 10^4}{0.4\pi W_a J K}$$

Very likely, after the permeability has been selected, a slight adjustment of the constraints to match the available core sizes and permeabilities must be made.

The outside diameter of the wound toroid, Fig. E-15 (less the outside wrapper), may be calculated from the following equation (assuming that one-half of the ID remains after winding):

$$\text{OD} = \sqrt{D_b^2\,(3/4) + D_c^2} \qquad (D = \text{diameter})$$

where
$$A_1 = \pi\,(R_b^2 - R_a^2)$$
$$A_1 = A_2$$
$$\text{core window ID} = D_a = 2R_a$$
$$\text{core ID} = D_b = 2R_b$$
$$\text{core OD} = D_c = 2R_c$$
$$\pi\,(R_b^2 - R_a^2) = \pi\,(R_d^2 - R_c^2)$$

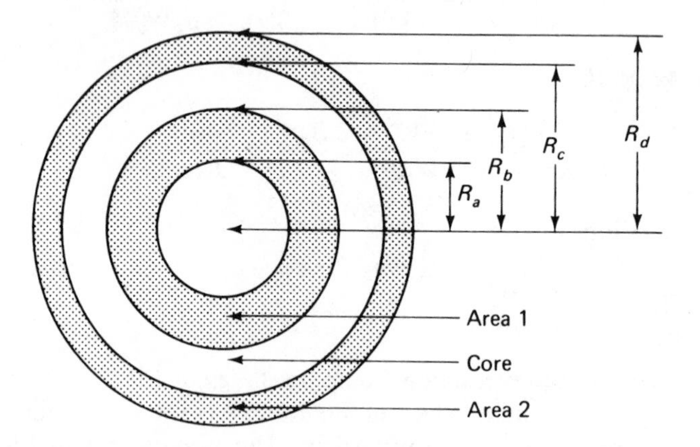

Figure E-15 Calculation of outside diameter of wound toroid.

$$R_b^2 - R_a^2 = R_d^2 - R_c^2$$

$$R_a = \frac{R_b}{2}$$

$$R_d^2 - R_c^2 = R_b^2 - \frac{R_b}{2}^2 = R_b^2 (1 - 1/4)$$

$$R_d^2 = R_b^2 (3/4) + R_c$$

$$R_d = R_b^2 (3/4) + R_c^2$$

$$\text{OD} = 2R_d$$

POWDERED-IRON TOROID CHARACTERISTICS

Table E-1 contains a comprehensive listing of powdered-iron toroid cores. The cores specified are available from Micrometals Corp. and Amidon Associates. The characteristics are the same for both brands, inclusive of the color codes. The dimensional data are given in English units. To convert English inches to millimeters, multiply inches by 25.4. These tabular data were provided through the courtesy of Micrometals Corp.

Dia. ± 0.005 inches Ht. ± 0.005 inches	Micrometals Number	L μ h 100 T	Typical Operating Frequency (MHz)	Color Code
ID 0.062 / 1.57; Std. ht. 0.050 / 1.27; OD 0.125 / 3.18	T12—1	48	2	Blue
	—2	20	8	Red
	—3	60	1	Gray
	—6	17	16	Yellow
	—7	18	14	White
	—10	12	40	Black
	—12	7.5	90	Green/white
	—15	50	2	Red/white
	—0	3.	150	Tan
B ht. 0.042 / 1.07	T12—2B	18.5	12	Red
	—6B	13.5	20	Yellow
	—10B	10	35	Black
ID 0.078 / 1.98; Std. ht. 0.060 / 1.52; OD 0.160 / 4.06	T16—1	44	1	Blue
	—2	22	9	Red
	—3	61	1	Gray
	—6	19	12	Yellow
	—10	13	30	Black
	—12	8	80	Green/white
	—15	55	1	Red/white
	—0	3	120	Tan

TABLE E-1

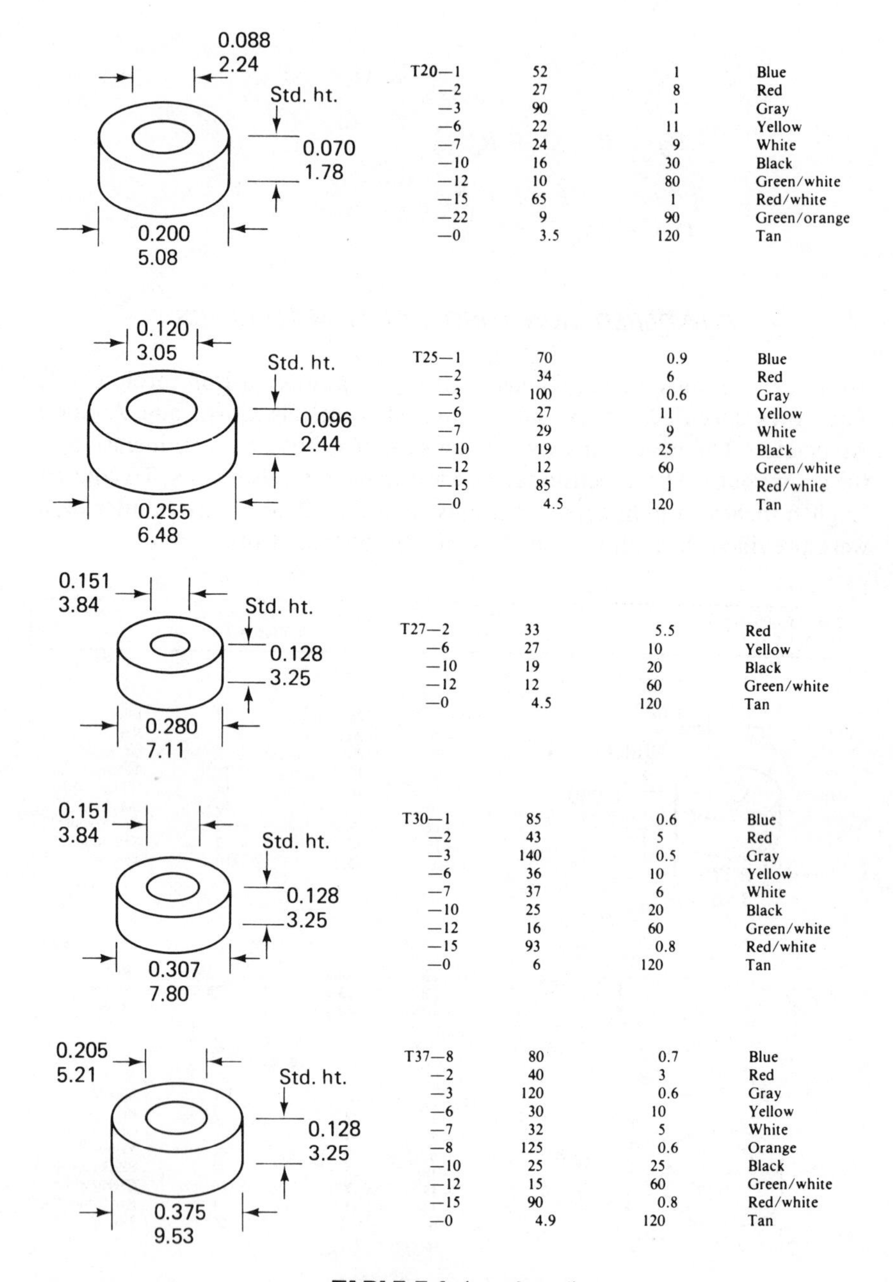

T20—1	52	1	Blue
—2	27	8	Red
—3	90	1	Gray
—6	22	11	Yellow
—7	24	9	White
—10	16	30	Black
—12	10	80	Green/white
—15	65	1	Red/white
—22	9	90	Green/orange
—0	3.5	120	Tan
T25—1	70	0.9	Blue
—2	34	6	Red
—3	100	0.6	Gray
—6	27	11	Yellow
—7	29	9	White
—10	19	25	Black
—12	12	60	Green/white
—15	85	1	Red/white
—0	4.5	120	Tan
T27—2	33	5.5	Red
—6	27	10	Yellow
—10	19	20	Black
—12	12	60	Green/white
—0	4.5	120	Tan
T30—1	85	0.6	Blue
—2	43	5	Red
—3	140	0.5	Gray
—6	36	10	Yellow
—7	37	6	White
—10	25	20	Black
—12	16	60	Green/white
—15	93	0.8	Red/white
—0	6	120	Tan
T37—8	80	0.7	Blue
—2	40	3	Red
—3	120	0.6	Gray
—6	30	10	Yellow
—7	32	5	White
—8	125	0.6	Orange
—10	25	25	Black
—12	15	60	Green/white
—15	90	0.8	Red/white
—0	4.9	120	Tan

TABLE E-1 (continued)

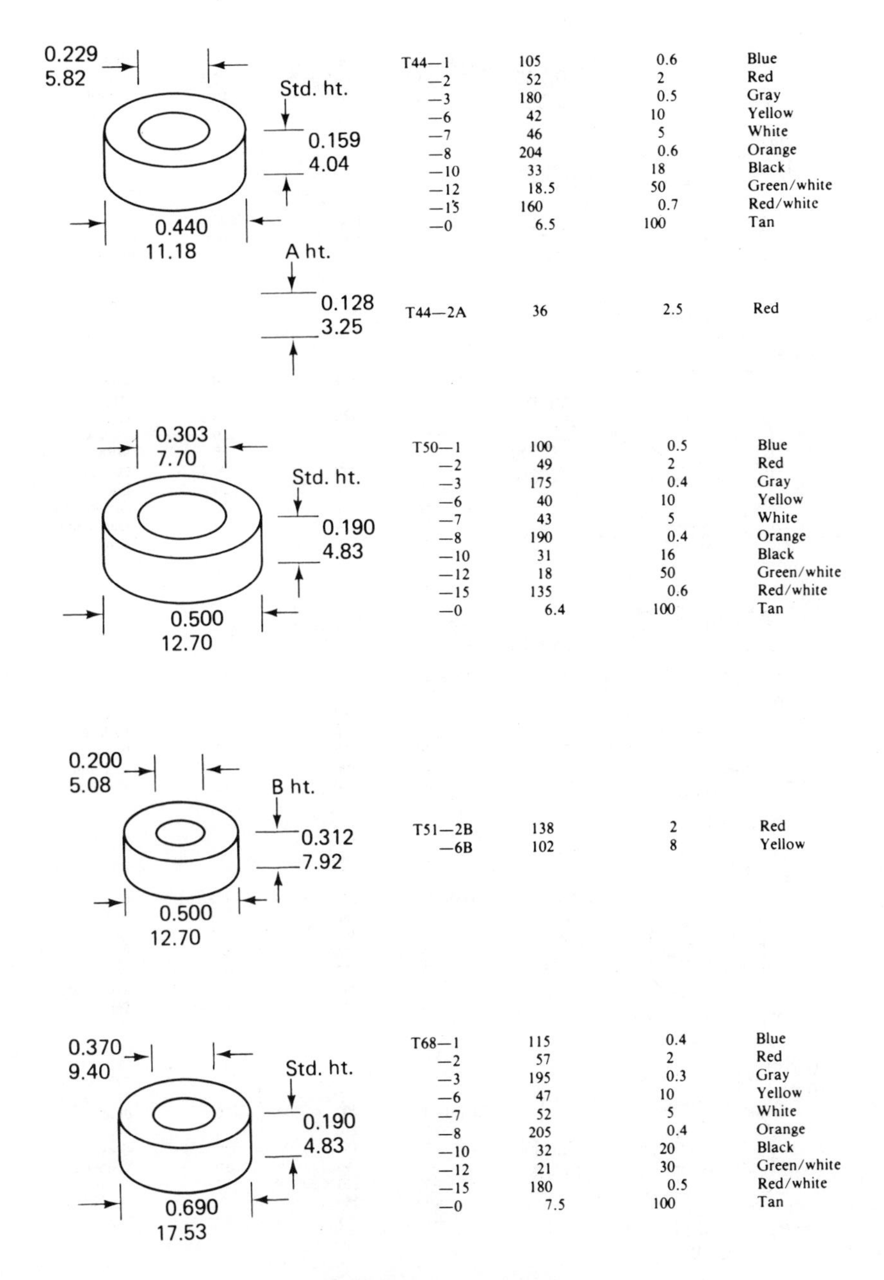

T44—1	105	0.6	Blue
—2	52	2	Red
—3	180	0.5	Gray
—6	42	10	Yellow
—7	46	5	White
—8	204	0.6	Orange
—10	33	18	Black
—12	18.5	50	Green/white
—15	160	0.7	Red/white
—0	6.5	100	Tan
T44—2A	36	2.5	Red
T50—1	100	0.5	Blue
—2	49	2	Red
—3	175	0.4	Gray
—6	40	10	Yellow
—7	43	5	White
—8	190	0.4	Orange
—10	31	16	Black
—12	18	50	Green/white
—15	135	0.6	Red/white
—0	6.4	100	Tan
T51—2B	138	2	Red
—6B	102	8	Yellow
T68—1	115	0.4	Blue
—2	57	2	Red
—3	195	0.3	Gray
—6	47	10	Yellow
—7	52	5	White
—8	205	0.4	Orange
—10	32	20	Black
—12	21	30	Green/white
—15	180	0.5	Red/white
—0	7.5	100	Tan

TABLE E-1 (continued

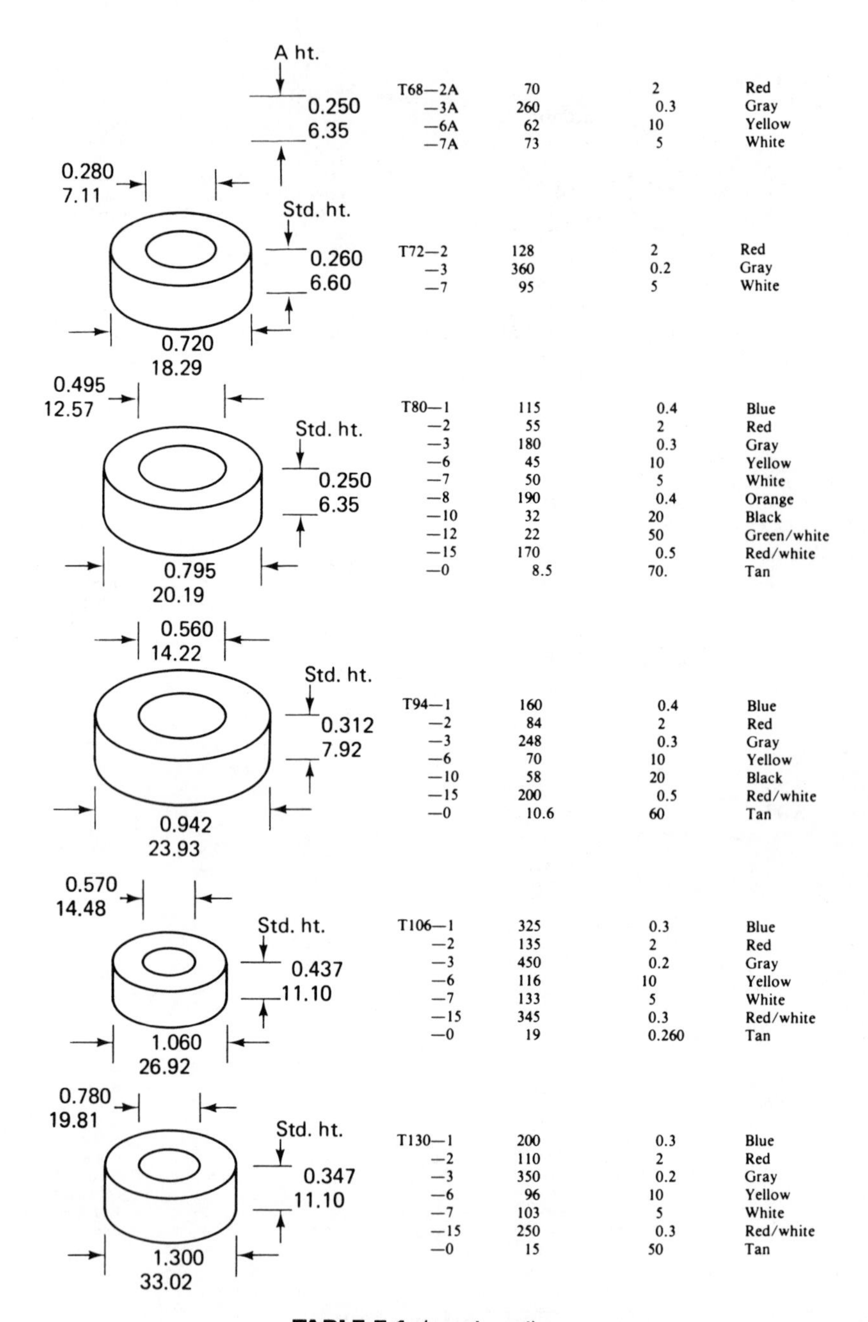

T68—2A	70	2	Red
—3A	260	0.3	Gray
—6A	62	10	Yellow
—7A	73	5	White
T72—2	128	2	Red
—3	360	0.2	Gray
—7	95	5	White
T80—1	115	0.4	Blue
—2	55	2	Red
—3	180	0.3	Gray
—6	45	10	Yellow
—7	50	5	White
—8	190	0.4	Orange
—10	32	20	Black
—12	22	50	Green/white
—15	170	0.5	Red/white
—0	8.5	70.	Tan
T94—1	160	0.4	Blue
—2	84	2	Red
—3	248	0.3	Gray
—6	70	10	Yellow
—10	58	20	Black
—15	200	0.5	Red/white
—0	10.6	60	Tan
T106—1	325	0.3	Blue
—2	135	2	Red
—3	450	0.2	Gray
—6	116	10	Yellow
—7	133	5	White
—15	345	0.3	Red/white
—0	19	0.260	Tan
T130—1	200	0.3	Blue
—2	110	2	Red
—3	350	0.2	Gray
—6	96	10	Yellow
—7	103	5	White
—15	250	0.3	Red/white
—0	15	50	Tan

TABLE E-1 (continued)

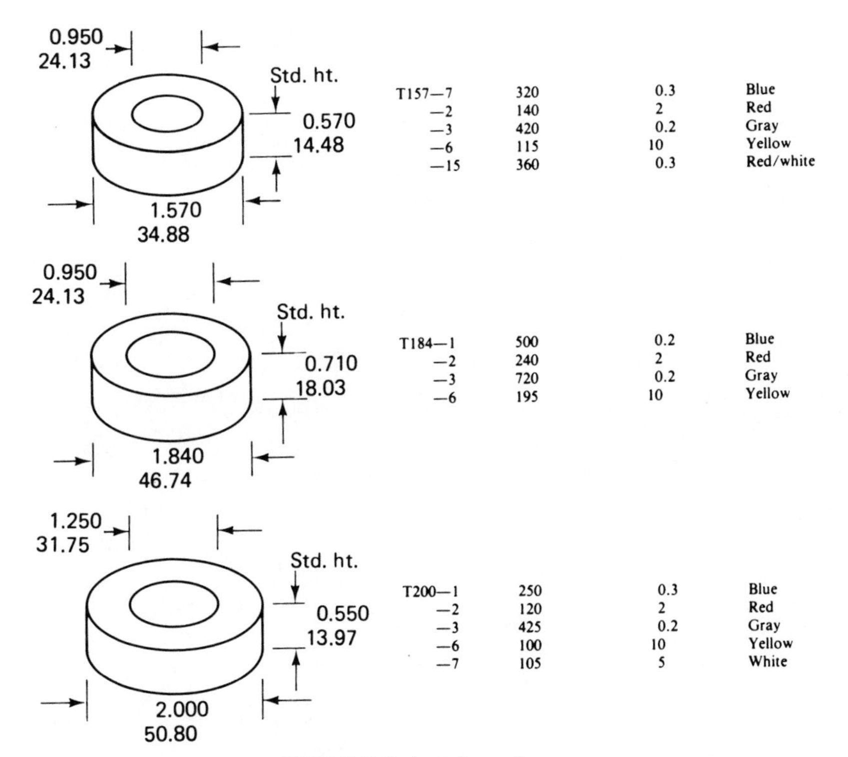

T157—7	320	0.3	Blue
—2	140	2	Red
—3	420	0.2	Gray
—6	115	10	Yellow
—15	360	0.3	Red/white
T184—1	500	0.2	Blue
—2	240	2	Red
—3	720	0.2	Gray
—6	195	10	Yellow
T200—1	250	0.3	Blue
—2	120	2	Red
—3	425	0.2	Gray
—6	100	10	Yellow
—7	105	5	White

TABLE E-1 (continued)

INDEX

N

P

Q

R

S

T

U

V

W